D1326012

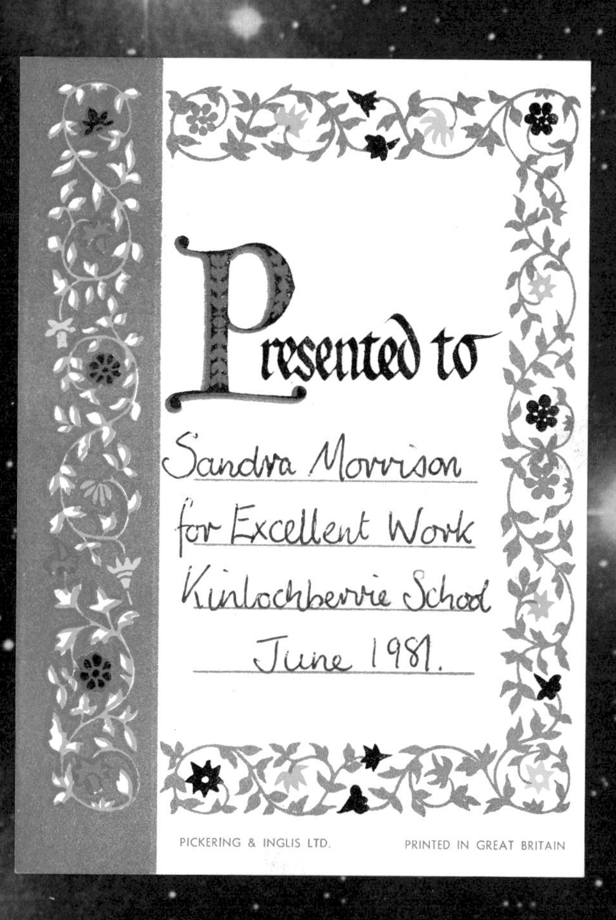
Presented to
Sandra Morrison
for Excellent Work
Kinlochbervie School
June 1981.
PICKERING & INGLIS LTD.
PRINTED IN GREAT BRITAIN

Hamlyn Encyclopedia of SPACE

Published 1981 by
The Hamlyn Publishing Group Limited
London · New York · Sydney · Toronto
Astronaut House, Feltham, Middlesex,
England

ISBN 0 600 38289 3

Printed in Spain

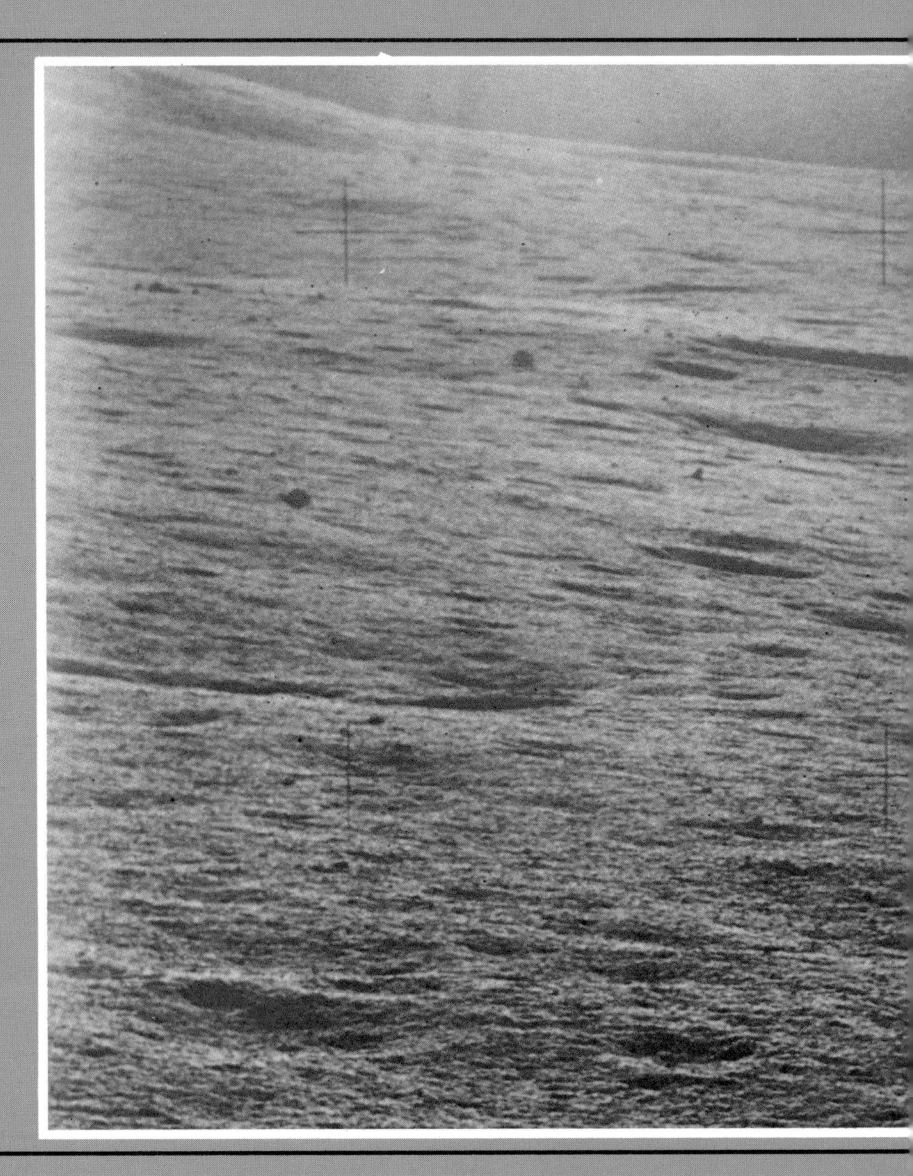

Hamlyn Encyclopedia of SPACE

Ian Ridpath

Hamlyn
London · New York · Sydney · Toronto

CONTENTS

ACKNOWLEDGEMENTS

Illustrations: Nigel Chamberlain, Jim Channell, Jon Davis, Phil Davis, Keith Duran, Tony Gibbons, Graham Gurr, Terry Hadley, Bill Hobson, Jack Pelling, Mike Tregenza, Brian Watson – through Linden Artists Limited.
Photographs: All were kindly supplied by NASA except for the following: Aerial Phenomena Research Organisation, Tucson, Arizona 152 above, 152 below; Associated Press, London 54 centre, 54 right; Aubrey Company/Paul N. Lazarus 111/AIP/London 147 top right; Stephen Benson, London 149, 150 top left, 150 top right; Copyright of the California Institute of Technology and the Carnegie Institution of Washington, reproduction by permission from the Hale Observatories, Pasadena, California 13 top right, 20 below, 21, 22, 23, 143; Canadian Government Photo Centre, Toronto 148; Columbia Pictures, Burbank, California 147 below; Department of the Environment, Crown Copyright (Tower of London) 32 top; Hamlyn Group Picture Library 14 top (Biblioteca Marucelliani, Florence), 14 bottom (Science Museum, London), 17, 18 right (inset), 32 bottom, 35 top right, 35 below; Imperial War Museum, London 39 below; Max Irvine 141; J. MacClancy 39 top; Novosti Press Agency, London back jacket middle right, 36, 47 top, 49 right, 50 centre, 51 below left, 52, 53, 54 top left, 59 top, 61 top left, 62 top left, 66 top left, 68, 90, 123 top right, 123 below left, 124 below left, 125 centre; Photri, Alexandria, Virginia 131 top (A. W. Richards); Popperfoto, London 125 right; Radio Times Library, London 140; Ian Ridpath 45 left, 50 top left; Ann Ronan Picture Library, Loughton, Essex 32 centre, 33, 34, 34-35, 35 top; Süddeutscher Verlag, Munich 39 left; Twentieth Century Fox Film Co. Ltd., London 144-145, 146-147.
Meteor crater picture (18) by courtesy of Meteor Crater Enterprises Inc., Winslow, Arizona.
Research at NASA by Frances C. Rowsell, Washington, D.C.

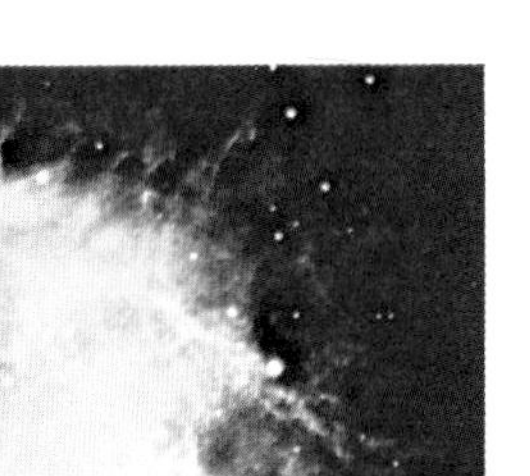

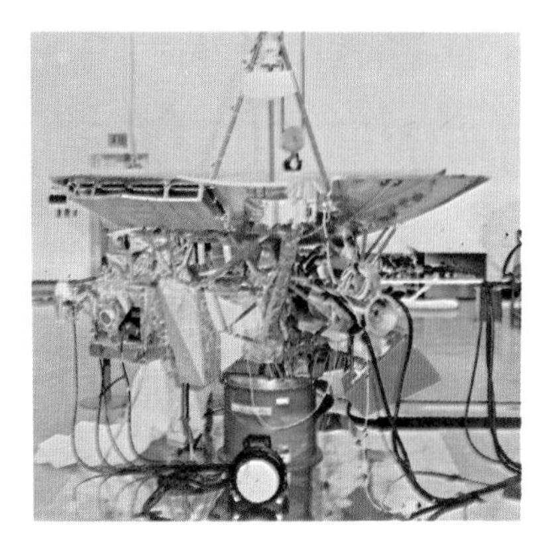

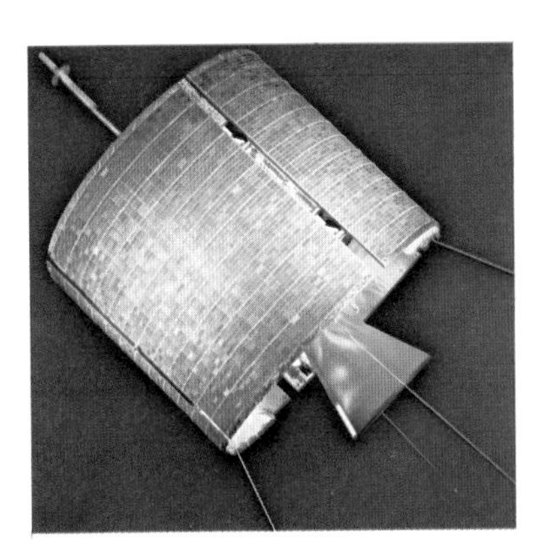

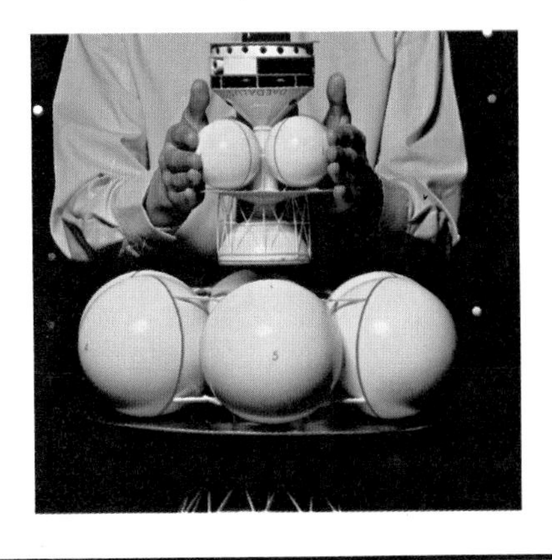

1 OUR PLACE IN SPACE

We live on a small planet called Earth, an island in a vast sea of space. Around us we can see other islands in space, separated from us by many millions of kilometres. These islands are the other planets that orbit the Sun. Nine planets, including Earth, are known to exist in the Sun's family, along with numerous pieces of rubble.

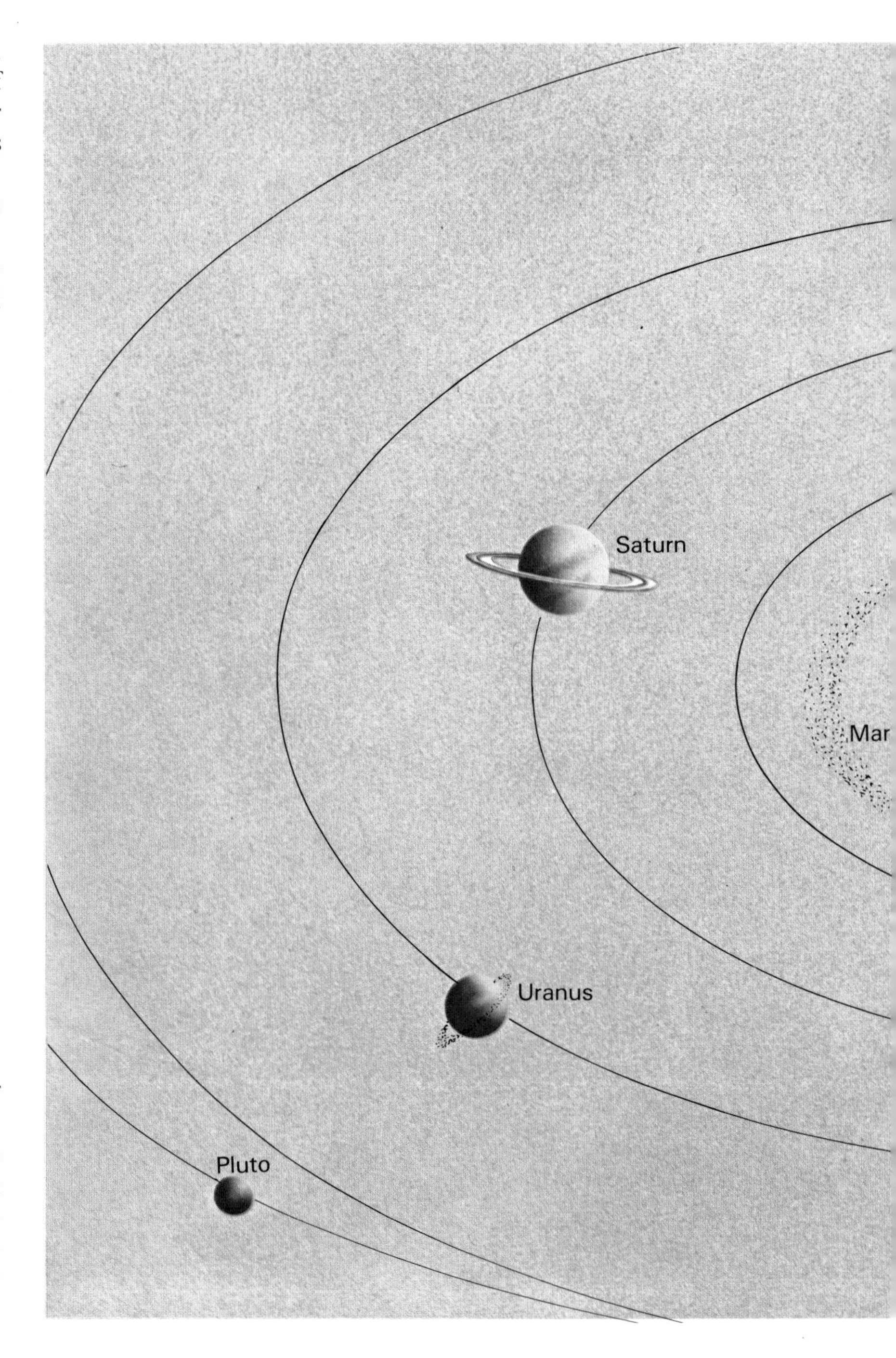

SEEING STARS

The Sun itself is a typical star, basically no different from the other stars we see in the night-time sky. These other stars look much smaller and fainter than the Sun simply because they are much further away. If we could travel to the stars, we would see that they are incandescent balls of gas like our own Sun. So far, space exploration has been confined to the nearest planets of our solar system. One day, however, we may be able to reach the stars and even set up bases there. Some stars may have planets on which other beings live and whom we may hope to contact by radio or by personal visits.

All the stars visible in the sky are part of a vast aggregation known as the Galaxy. The Galaxy contains hundreds of thousands of stars; we see only the nearest and brightest of them at night. The Galaxy is shaped like a giant catherine-wheel. Most stars are congregated along the plane of the wheel, forming the faint misty band we see as the Milky Way.

The name Milky Way is often given to our entire Galaxy.

But, beyond the Milky Way, there are countless other galaxies, revealed by large telescopes. Galaxies stretch into space as far as we can see; every time a bigger telescope is built, more galaxies can be seen. All the galaxies and the space around them make up the Universe. By studying galaxies in the most distant parts of space, astronomers hope to understand how the Universe began and how it has evolved.

OUR NEIGHBOURS IN SPACE
Exploration of space by satellites and probes is helping scientists learn more about our neighbour worlds in the solar system, and about the Universe as a whole. This book tells you about the discoveries that have been made, the way in which space is being used to help mankind, and possible developments in the future–many of which you will see come true in your own lifetime.

Nine planets orbit the Sun, along with smaller pieces of debris, making up the solar system.

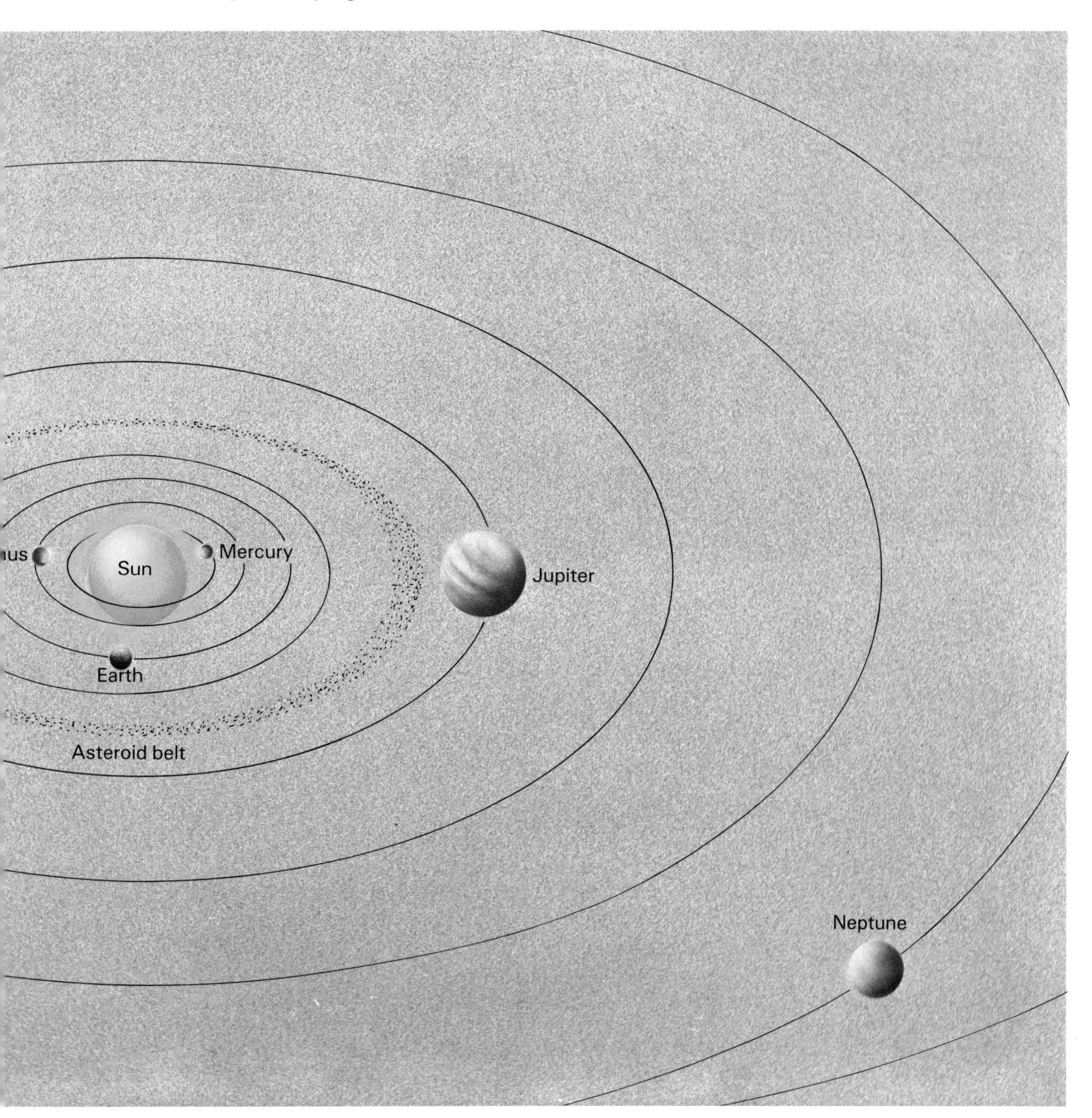

Brilliant star fields, dark clouds of dust, and glowing red gas clouds in the Milky Way in Sagittarius.

Mercury

First, let's take a closer look at our neighbour worlds in space, starting with the closest planet to the Sun, Mercury. This is a small, rocky body only 50 per cent larger than our own Moon, with no air or water. (Complete facts and figures about the planets are shown in the table on page 10).

Mercury's surface, like that of our Moon, is scarred with craters. So near is the planet to the Sun that its daytime side is heated to a temperature of 350°C (degrees Centigrade), hot enough to melt tin and lead. But, because there is no atmosphere to hold the heat in, temperatures on Mercury's night side fall to 170°C below freezing.

Mercury is a very inhospitable world indeed, and not a place likely to be visited by astronauts. It is difficult to see from Earth and the only clear views of its surface have come from a space probe.

Spectacular craters and ridges on the surface of Mercury, photographed by the Mariner 10 space probe.

Venus

The second-closest planet to the Sun is Venus, which is of similar size to Earth but otherwise very different. Venus at its brightest is an unmistakable sight; it blazes more brilliantly in the skies of Earth than any object other than the Sun and Moon. There are two reasons for its brilliance, one being that it can come closer to Earth than any other planet–as close as 38 million km (24 million miles). The second reason is that Venus is covered with an unbroken blanket of clouds, which reflects most of the Sun's light.

These clouds prevent astronomers from seeing the surface of Venus, so most of what we know about the planet has come from space-probe exploration. Underneath its clouds, Venus has a dense atmosphere but, instead of being mostly nitrogen and oxygen as is the atmosphere of Earth, it is made largely of unbreathable carbon dioxide gas. Carbon dioxide traps heat from the Sun like a blanket, with the result that tem-

Above: Crescent Earth as seen by the crew of Apollo 17 orbiting the Moon, and, *inset*, the full Earth showing Africa and Antarctica.
Left: Dramatic, swirling clouds of Venus photographed by Mariner 10.

peratures at the surface of the planet are like a furnace, reaching nearly 500°C. In view of these temperatures it is no surprise that there is scarcely any water on Venus. Not even the clouds are made of water vapour; instead, they consist of concentrated sulphuric acid. Despite its heavenly name, Venus has turned out to be more like hell.

One odd fact about Venus is that it rotates on its axis from east to west instead of from west to east like the other planets, and it takes longer to spin on its axis than it does to orbit the Sun. The reason for this remarkable state of affairs remains unexplained.

Earth

Third in line from the Sun is our home planet, Earth, which also appears quite bright in space

PLANETS – DATA

	Planet	Average distance from Sun (million km, followed by million miles in brackets)	Diameter (km, followed by miles in brackets)	Circles Sun in	Turns on axis in	Known moons
1	**Mercury**	57.9 (36)	4880 (2950)	88 days	58.7 days	0
2	**Venus**	108.2 (67.2)	12 100 (7500)	224.7 days	243 days	0
3	**Earth**	149.6 (93)	12 756 (7920)	365.25 days	23.56 hours	1
4	**Mars**	227.9 (141.5)	6790 (4200)	687 days	24.37 hours	2
5	**Jupiter**	778.3 (483.6)	142 200 (88 700)	11.9 years	9.50 hours	14
6	**Saturn**	1427 (886.6)	119 300 (75 100)	29.5 years	10.14 hours	10
7	**Uranus**	2870 (1782)	51 800 (32 000)	84 years	20 hours*	5
8	**Neptune**	4497 (2794)	49 500 (30 000)	165 years	18.5 hours*	2
9	**Pluto**	5900 (3666)	3000* (1865)*	248 years	153 hours	1

Notes: * means data uncertain

5
6

Left: The Moon, showing parts of its Earth-facing and far hemispheres, photographed from Apollo 11.
Above: Rugged mountains mark the north-west rim of the lunar crater named King. Part of the crater's complex central peak is seen at top right of photograph, taken by Apollo 16 astronauts orbiting the Moon.

because of the extensive clouds in its atmosphere. The clouds of Earth, unlike those of Venus, are made of water vapour, and there is, of course, abundant water in the seas, too. We might term Earth the blue and white planet from the colour of its seas and clouds, were we to view it from far off in space. Satellites, of course, are helping us learn a great deal more about our own planet and its environment in space.

Earth lies at just the right distance from the Sun for water to exist in liquid form. Were Earth much closer to the Sun it would be so hot that water would boil, as has happened on Venus. If it were further away water would freeze, as happens on Mars. This abundance of liquid water is the major reason why Earth is favoured for life. Life is believed to have originated in the seas, and water is vital to the continued survival of all forms of life on Earth.

Earth is remarkable in that it has a moon over one-quarter its own diameter. The Moon's precise diameter is 3476 km (2160 miles). The Moon lies 384 400 km (239 000 miles) from us. As seen from afar, Earth and the Moon would probably be termed a double planet. The Moon is important because it raises tides in the oceans of Earth and provides light at night. It is also by far the closest natural body to us in space and so has become a target for probes and manned landings. It is an airless and waterless world that has been heavily battered by meteorites, producing the cratered terrain visible through even a simple pair of binoculars.

Mars

Beyond Earth we come to Mars, commonly known as the red planet because of the distinctive colour of its deserts. Mars is half the size of Earth and has a thin atmosphere of carbon dioxide in which clouds form. The planet was long considered a possible haven of life, both in science fiction and science fact.

Controversy raged at the turn of the century about the supposed 'canals' of Mars–straight, dark markings which some observers, notably the American Percival Lowell (1855–1916), claimed were artificial waterways dug by Martian beings. Most other astronomers could not see the canals, and space probes have now disproved their existence, but until recently it was thought possible that some form of micro-organism might exist on Mars. Alas, the Viking space probes which landed on the planet in 1976 failed to find any definite signs of life.

In the light of what we now know about conditions on Mars, this lack of life is not altogether surprising. Its scant atmosphere,

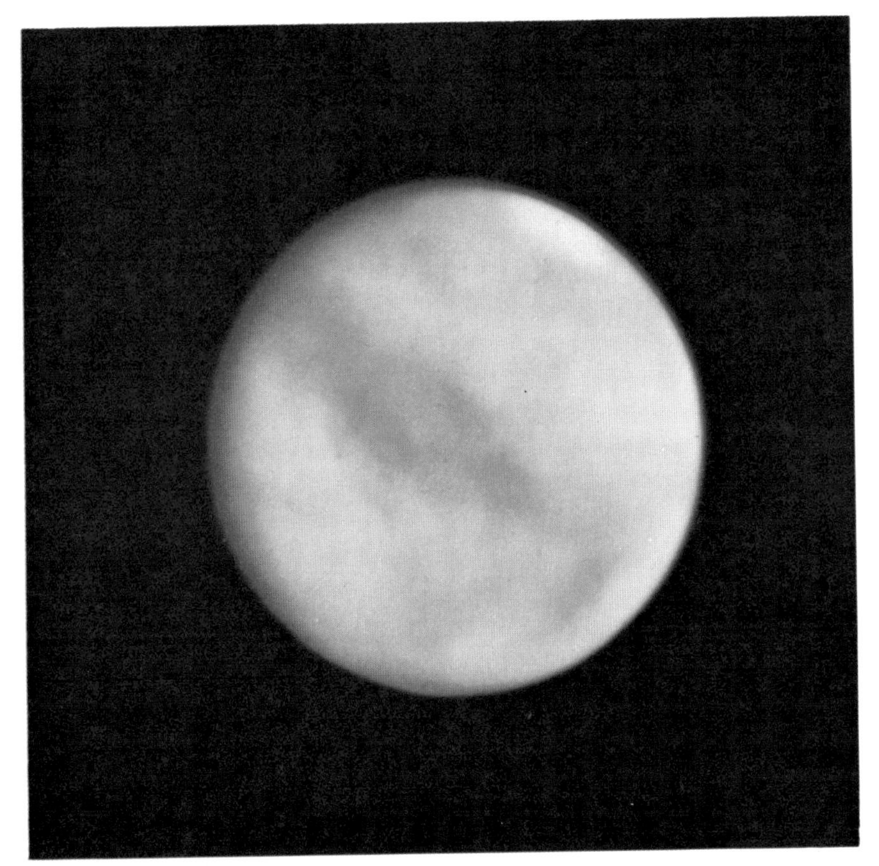

Above: Craters on Mars near the planet's south polar cap photographed by the Mariner 7 space probe. White deposits of frost fill the craters.
Right: Mars as seen from Earth through a telescope, showing its orange colour, some dark markings, and a white polar cap.
Below right: Jupiter, the giant planet of the solar system, photographed by Voyager 1. Note the red spot at lower left, and one of Jupiter's satellites, Io, against the planet's disk at centre right of the photograph.

combined with its distance from the Sun, means that temperatures on Mars never rise above the freezing point of water. Therefore Mars has no liquid water, but it does have polar caps made of ice, and there is probably also a layer of permafrost under the whole of the planet's surface. Parts of the surface look like the Moon, peppered with large craters apparently formed by meteorites, but there are also giant volcanoes and valleys which prove there could be geological activity on the planet. Astronauts will eventually go to Mars, though probably not until next century.

Jupiter

Jupiter, next in line beyond Mars, is the largest planet of the solar system, weighing two and a half times as much as all the other planets rolled together. Jupiter is a giant ball of gas, mostly hydrogen and helium, 11 times the

Top: Galileo, the Italian scientist who discovered the four main moons of Jupiter in 1610.
Above: Galileo's tiny telescope.

diameter of Earth. Colourful bands of clouds swirl around it, producing belts and streaks visible in the smallest telescope. One prominent feature of the clouds is the red spot, a giant, red-coloured oval that has remained for centuries. The red spot is so big it could swallow several Earths.

Astronauts could not land on Jupiter because there is no solid surface to speak of. Instead, the gases of which Jupiter is made get denser and denser towards the planet's centre. Any descending space probe would continue falling until it was crushed by the pressure.

Binoculars held steadily will reveal up to four faint pinpoints of light either side of Jupiter. These are its four brightest moons, discovered by Galileo in 1610. They change position from night to night as they orbit the planet. A total of 14 moons of Jupiter have now been reported, more than for any other planet, and there may be others too faint to see.

Saturn

Saturn, sixth in line from the Sun, is perhaps the most beautiful planet of all. It is a ball of gas similar in nature to Jupiter, although somewhat smaller. Its most striking feature is a set of brilliant rings, made of countless particles all moving in orbits around the planet like miniature moons. Small telescopes show the rings, which stretch approximately 275 000 km (170 000 miles) from rim to rim.

The ring particles, probably about the size of bricks, are coated with frozen gas. It is this which makes them so bright. The rings are believed to be the building blocks of a moon which never formed.

Saturn alone has at least 10 moons, the largest of which, Titan, is bigger than the planet Mercury and has an atmosphere of its own.

Uranus

Beyond Saturn, far from the Sun, the solar system becomes darker and more mysterious. In fact until 1781 it was believed the solar system stopped at Saturn. But, in that year, the astronomer Sir William Herschel discovered a faint planet, which was named Uranus. It was another gaseous giant, about half the size of Saturn, yet around four times larger than Earth.

Uranus turned out to be a strange planet, with an axis inclined at more than 90° to the vertical, as though the planet had been toppled on its side by some catastrophic collision long ago. In 1977, astronomers discovered that Uranus has rings, much darker than the rings of Saturn. Nine rings are believed to exist, and may represent the fragments of a former moon that strayed too close to the planet and was broken up by tidal forces.

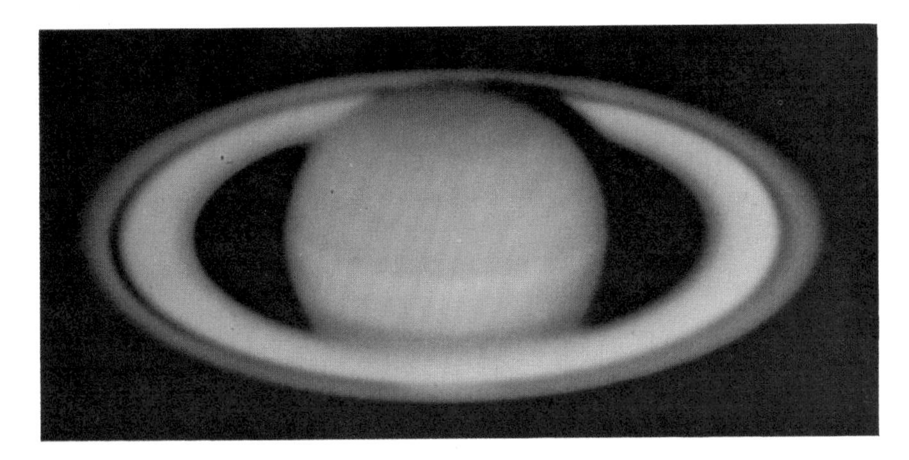

Left: Saturn as seen through a telescope. Many people consider Saturn the most beautiful planet in the solar system because of its system of bright rings, composed of tiny particles in orbit around it.
Below: Artist's impression of the rings around the planet Uranus. The rings of Uranus, discovered in 1977, are much fainter than those of Saturn. Nine rings are now known to exist around this planet.
Bottom: Seen through a telescope, the remote planet Neptune shows a greenish-coloured, featureless disk, as in this artist's impression.

Neptune

Following its discovery, Uranus was seen to be slowly straying off its predicted course, as though it were being tugged by the gravity of some other, still unknown, planet. Mathematicians set about calculating the likely position of this new body and, in 1845, the planet Neptune was discovered close to the predicted spot. Neptune turned out to be a near twin of Uranus, though without being tilted on its side. One of Neptune's two moons, Triton, may be the largest moon in the solar system, with a diameter of about 6000 km (3730 miles) according to some measurements. Neptune may have rings like those of Uranus, but none have yet been found.

Once astronomers had discovered Neptune, they began to wonder if still more planets lay in the darkness beyond. The American astronomer Percival Lowell, of Martian canal fame, began a major search for a new body he called planet X. He died before the search was completed but in 1930, at Lowell's observatory in Arizona, the planet Pluto was eventually found.

Pluto

Pluto turned out to be a surprise. It was not a giant gaseous body like Uranus and Neptune, as had been expected, but was small and

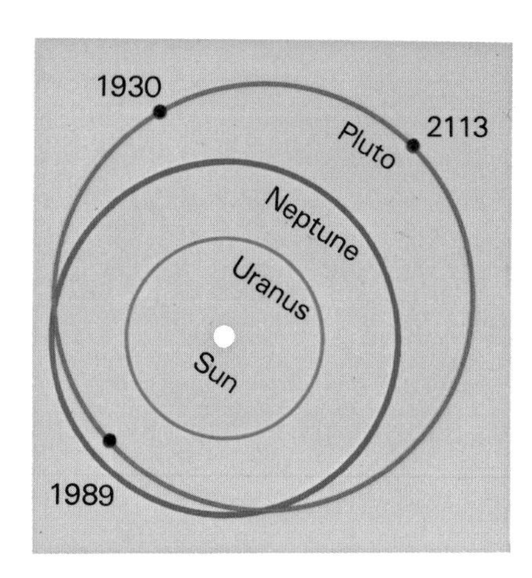

Pluto, discovered in 1930, has an elliptical orbit that crosses the path of Neptune. Pluto will be at its closest to the Sun in 1989, and at its furthest away in 2113.

faint. The latest measurements show that Pluto is the smallest planet in the solar system, and is smaller even than our own Moon. One theory is that Pluto is not a true planet at all, but an escaped satellite of Neptune. In 1978, astronomers discovered that Pluto has a moon.

Pluto is not at present the outermost planet of the solar system. Although Pluto's *average* distance from the Sun is greater than that of any other planet, its orbit is so eccentric it can at times come closer to the Sun than Neptune, and this has now happened. From January 1979 until March 1999, Neptune is the furthest planet from the Sun. Then Pluto re-claims its rightful position at the edge of the solar system. Pluto is the only planet to cross another planet's orbit.

COMETS, ASTEROIDS AND METEORITES

Astronomers have looked for more planets beyond Pluto, but nothing has been found. Most people now agree that the planets end with Neptune and Pluto. However, there are believed to be plenty of the ghostly wanderers called comets in the dark outer reaches of the solar system.

Comets, which resemble dirty snowballs, are bags of dust and gas travelling on elongated orbits around the Sun. Occasionally they swoop into the inner part of the solar system where they become bright enough for us to see.

Something must also be said about minor planets, also known as asteroids. At least 100 000 are estimated to be visible in large telescopes, mostly orbiting in a swarm between Mars and Jupiter, though some stray more widely through the solar system. The largest asteroid, Ceres, has a diameter of no more than 1000 km (620 miles) and all the asteroids rolled together would not make one body the size of Earth. Probably they are rubble left over from the formation of the solar system.

Sometimes, lumps of rock and metal from space come hurtling into Earth's atmosphere. These are meteorites, which are believed to be fragments of asteroids. Impacts by asteroids and meteorites probably caused most of the craters on the planets. There are a few known meteorite craters on Earth, notably the one near Winslow, Arizona, but fortunately the Earth's atmosphere protects us from too many collisions with interplanetary wanderers.

This survey of the solar system has necessarily been brief, but we shall return to the planets and their exploration by space probes in Chapter 6. For the moment, let

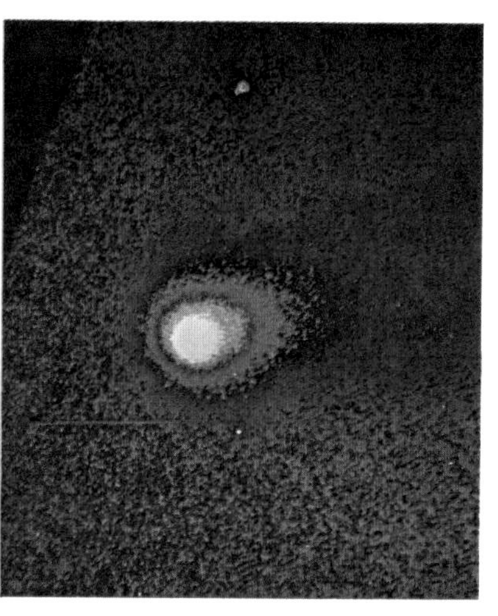

Right: The glowing head of comet Kohoutek photographed in ultraviolet light from space by the Skylab astronauts. The colours have been added artificially to show different levels of brightness. *Below:* Comet Kohoutek photographed by astronomers in Arizona after it rounded the sun in January 1974.

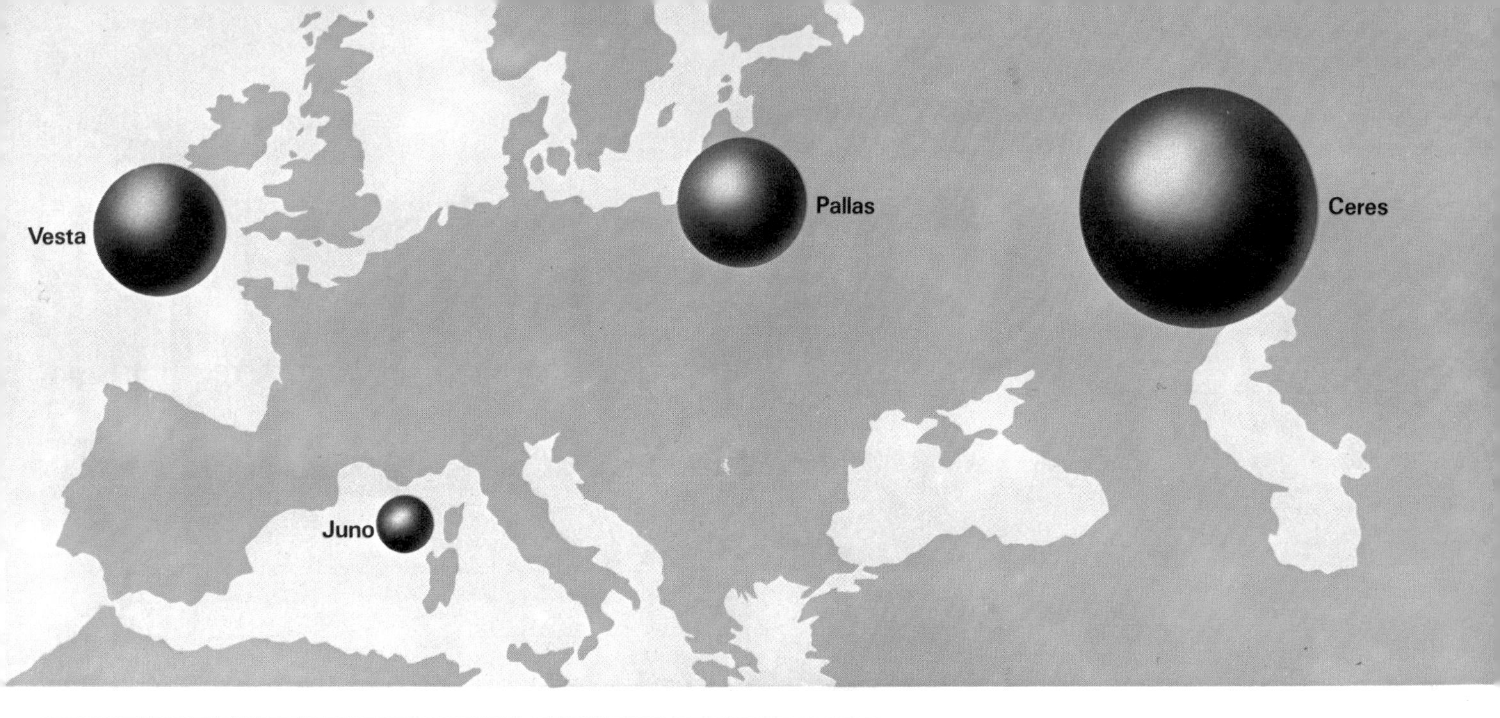

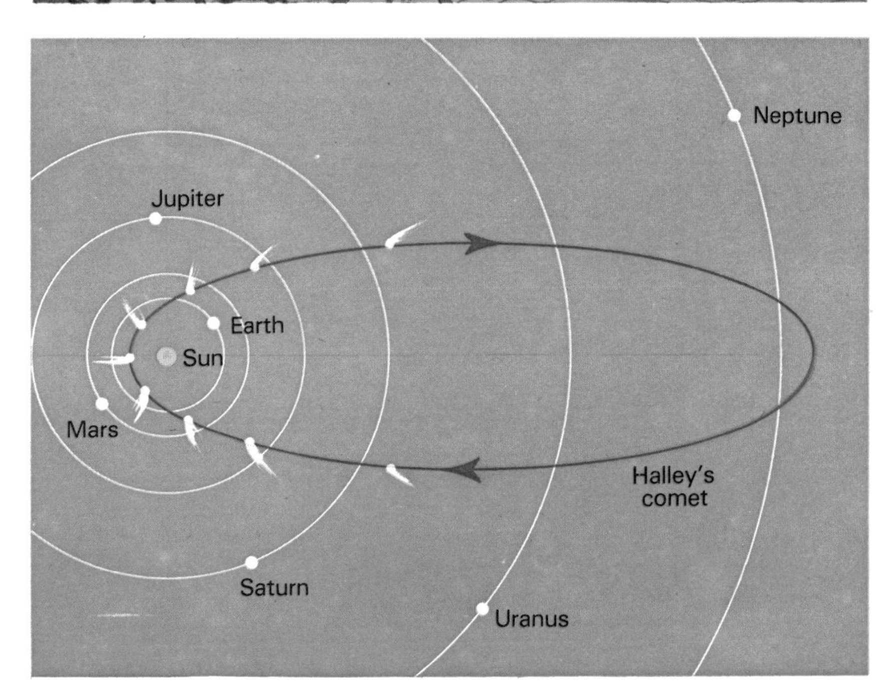

Above: Sizes of the largest asteroids compared with Europe. Ceres, the largest asteroid of all, is nearly 1000 km (625 miles) in diameter.
Left: Halley's comet, shown on the Bayeux tapestry, was regarded as an omen of the Norman defeat of King Harold when it appeared in 1066.
Below left: Halley's comet has a highly elliptical orbit around the Sun. It takes about 76 years to orbit the Sun.

us turn our attention to the object at the heart of the solar system–the Sun.

THE SUN, A GLOWING BALL OF GAS

The Sun is a glowing ball of gas 1 392 000 km (865 000 miles) in diameter. About 80 per cent of it is hydrogen and almost all the rest is helium, with only a sprinkling of other substances. A distance of 150 million km (93 million miles) separates the Sun from Earth which, in astronomical terms, is quite close. Light takes only 8.3 minutes to reach us from the Sun as against more than four years from the nearest other star, Alpha Centauri. Astronomers say that Alpha Centauri is four light years away. The light year is the basic unit of astronomical distances. It is equivalent to 9.5 million million km (six million million miles).

Whereas planets are dark bodies that shine only by reflecting

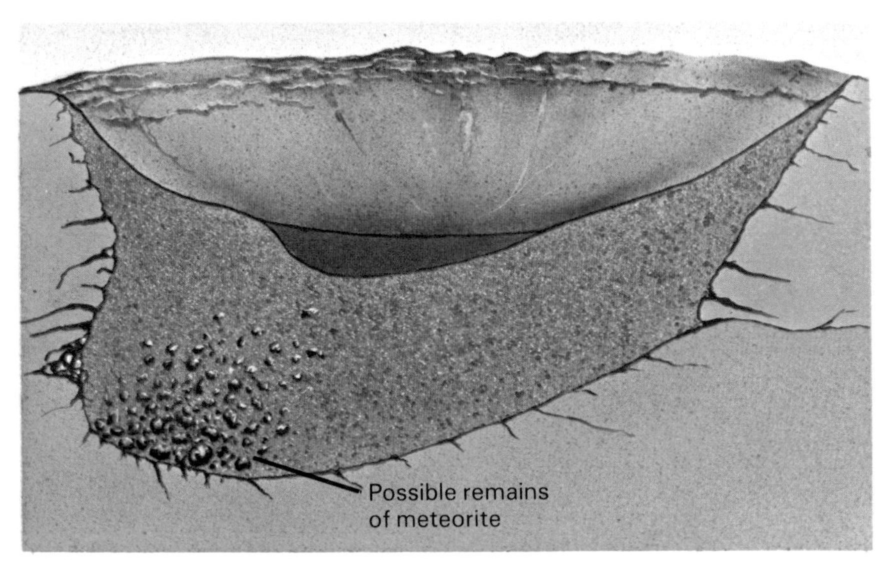

Left: Cross-section of the Arizona meteorite crater. Remains of the meteorite which caused the crater are believed to be buried deep underground.
Bottom: Meteorite crater near Winslow, Arizona. One km (0.62 of a mile) in diameter, it was blasted out by the impact of a giant meteorite 50 000 years or so ago.
Below: Local villagers gather round a meteorite that fell to Earth in Mexico.

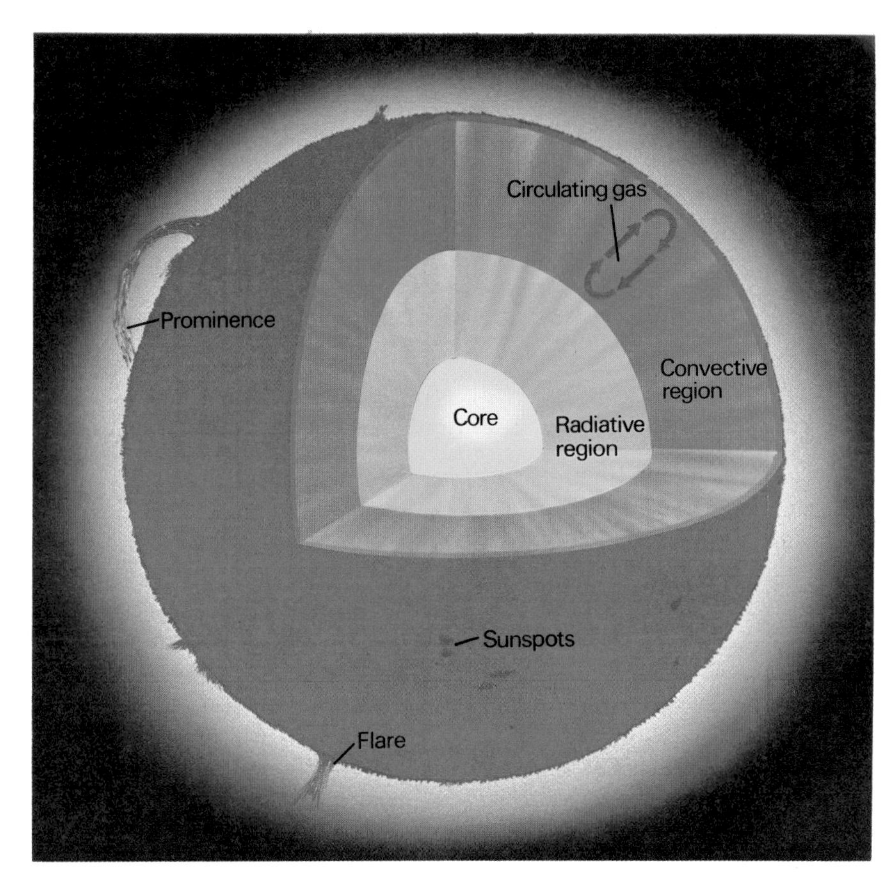

sunlight, stars produce their own light and heat through nuclear reactions. There is a minimum limit to the mass of a star. It must be at least 0.06 the mass of the Sun (60 times the mass of Jupiter), or else conditions at its centre do not become extreme enough for nuclear reactions to begin. Therefore if Jupiter had been 60 times heavier, it would have become a small star.

The centre of the Sun is a giant nuclear furnace. There, conditions of temperature and pressure are so extreme that atoms of hydrogen fuse together to form atoms of the heavier element helium. As this happens, energy is released– energy which keeps the Sun glowing, and which provides us with the light and heat essential to life. This same process of fusing hydrogen into helium occurs in a nuclear bomb. But, thankfully, the Sun does not fly apart in a vast nuclear explosion because its own gravity is strong enough to hold it together.

Processes that occur on the Sun are of great interest, because they affect all life on Earth. Every 11 years or so, the Sun goes through a cycle in which storms and eruptions on its surface become more frequent, and its surface is marked by dark patches known as sunspots. Atomic particles are thrown off into space during solar storms, and they can affect the outer layers of Earth, causing effects such as radio blackouts.

The Sun's 11-year cycle of activity seems to affect Earth's

Left: A cutaway diagram of the Sun. Energy is generated by nuclear reactions at the Sun's core and flows outwards, completing its journey in vast convection cells of gas.
Below: A massive eruption of helium gas stretching 800000 km (half a million miles) into space from the surface of the Sun, photographed through telescopes aboard the Skylab space station.

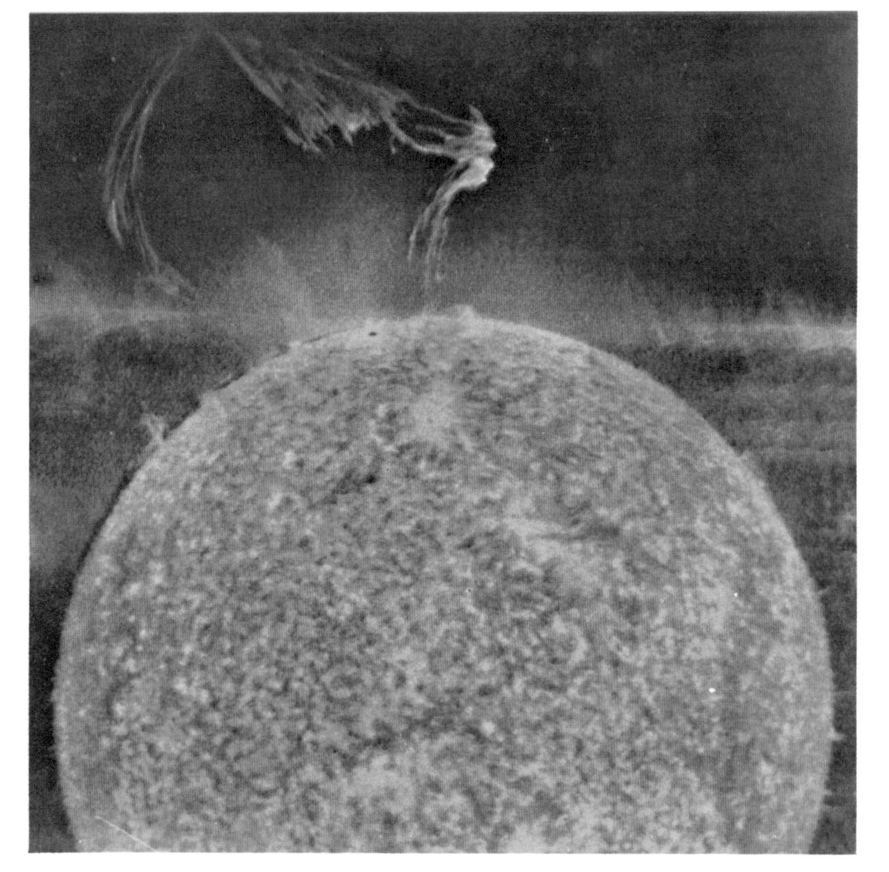

weather in ways that are not understood. Longer-term changes on the Sun may have significant effects on Earth's overall climate, perhaps even leading to new ice ages. Since minor changes in rainfall and temperature can seriously affect our energy needs and the amount of food we can grow to feed a hungry world, these fluctuations of the Sun are of vital importance. Satellites and probes are being used to monitor what happens on the Sun. One day it may be possible to predict what changes are likely to occur.

Some stars are bigger and hotter than the Sun while others are smaller and cooler. But, on the whole, our Sun is pretty typical of stars, so that by learning more about the history and nature of the Sun we are learning more about stars in general.

From Just a Cloud in Space . . .

How do stars form? According to current theories, a star is born from a giant cloud of gas and dust, of which many still exist in the Galaxy. One famous example is the great nebula in Orion, visible to the naked eye and a magnificent sight through binoculars or a small telescope. At the heart of the nebula are several new-born stars and it is their light which causes the nebula to glow brightly.

Our Sun is believed to have formed from just such a cloud in space, about 4600 million years ago. As the cloud began to shrink under its own gravity, it broke up into tiny blobs which gradually became smaller and therefore hotter. Eventually, conditions at the centre became so extreme that nuclear reactions 'switched on', turning the object into a true star; these reactions continue to power the Sun today. Probably the Sun formed as a member of a cluster of stars, the other members of which have since drifted away.

One famous star cluster is the Pleiades in Taurus, which contains about 200 stars which have formed in the past 50 million years or so– quite recently in astronomical terms. They have not yet had time to drift apart. You can see the brightest of the Pleiades with the naked eye or through binoculars. Long-exposure photographs with telescopes show that the stars are

Left: An eruption at the edge of the Sun photographed from Skylab. Colours were added electronically.

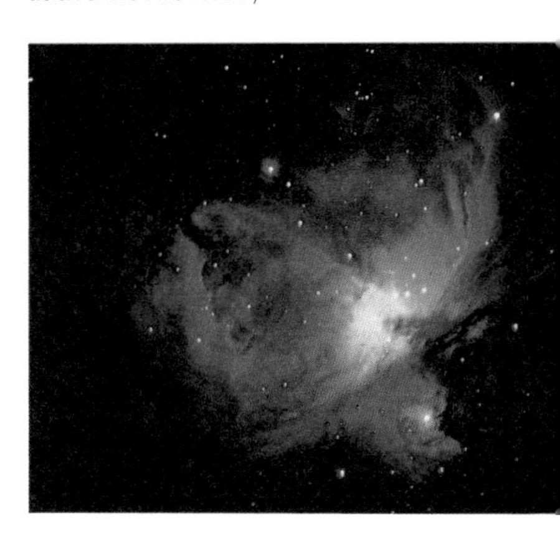

Above: The Orion nebula, a cloud of gas in space, is the birthplace of stars.
Below: The horsehead nebula, an area of dark dust seen silhouetted against glowing gas.

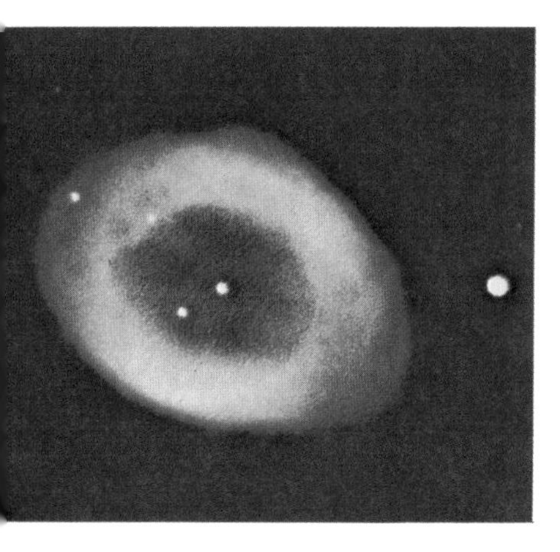

Above; The ring nebula in Lyra, a celestial smoke ring which marks the end of life of a star like the Sun.
Right: The Pleiades in Taurus, a cluster of young stars born from a gas cloud in space.

still surrounded by traces of the nebula from which they formed.

Eventually, though, a star must run out of fuel at its centre. When this happens, the nuclear reactions move further out into the layers surrounding the core. The star gets hotter because it is releasing more energy. It swells up in size and becomes what is known as a red giant, perhaps 100 times its original diameter.

When this happens to the Sun, as it inevitably must, all life on Earth will shrivel to a cinder. Fortunately, though, the Sun is not scheduled to become a red giant for about another 5000 million years–so we have plenty of time!

Supernova

Once a star reaches the red giant stage it is near the end of its life. What happens next depends on how massive the star is. A star like the Sun would lose its distended outer layers to form a celestial smoke ring. At the centre of this smoke ring is the white-hot core of the former star, which remains as a tiny white dwarf star.

But in stars several times more massive than the Sun, conditions of temperature and pressure are more extreme. When such stars reach the ends of their lives, a series of runaway nuclear reactions takes place, making the star unstable. It erupts in a nuclear holocaust known as a supernova, becoming millions of times brighter for a period of a few months.

Oriental astronomers in AD 1054 saw just such a stellar eruption in the constellation of Taurus. The shattered remains of that star form the object known as the Crab nebula, visible in telescopes on Earth today.

A supernova explosion does not always completely destroy a star. Often the core of the star is left behind as a tiny, compressed object known as a neutron star. Neutron stars are so called because the protons and electrons of which they are made have been squeezed together to form the atomic particles known as neutrons. Neutron stars are extremely dense. Whereas a white dwarf is about the size of Earth, a neutron star contains as much matter squeezed into a ball only 20 km

The Crab nebula in Taurus is the remains of a star which exploded in an eruption known as a supernova.

Right: The North American nebula, a glowing cloud of pink gas, lies in a rich part of the Milky Way in Cygnus. The nebula takes its name from the fact that its shape resembles the continent of North America. Turn the page on its side and you will see this resemblance.

Black holes: when a large and heavy star explodes at the end of its life, it can leave behind a small, dense object with such a strong gravitational pull that nothing can escape, not even light. Therefore the object is invisible - a black hole.

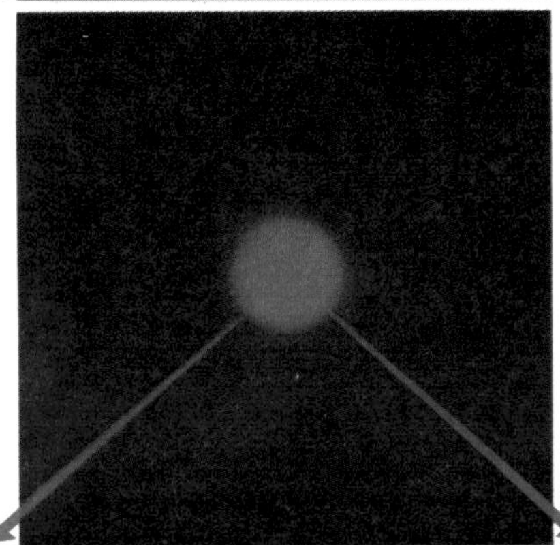

(12.5 miles) or so in diameter. A spoonful of white dwarf material would weigh several tonnes; a spoonful of material from a neutron star would weigh perhaps 1000 million tonnes.

Pulsars

Since neutron stars are so small, they are also extremely faint. Only two of them have ever been seen visually–the one at the centre of the Crab nebula, and one in the constellation of Vela, which is, incidentally, the faintest object ever identified by astronomers. But it turns out that neutron stars can be detected in other ways, particularly with radio waves, and this is how they were first found.

In 1967, radio astronomers in Cambridge, England, picked up rapid radio pulses coming from a number of directions in space. They called these objects pulsars. As more pulsars were discovered it became clear they were actually rapidly rotating neutron stars which gave out a flash of radiation like a lighthouse beam each time they turned.

Typical pulsars flash once every second or two; the fastest of all, at the centre of the Crab nebula, flashes 30 times a second. Once this pulsar's radio pulses were detected it was also seen flashing optically. Pulses from this object have also been detected at other wavelengths, notably X-rays, by space satellites. Many other radio pulsars have also been detected at X-ray wavelengths.

Not all neutron stars are pulsars. These radio-quiet neutron stars would be undetectable were it not for the fact that many are found orbiting normal stars. Gas from the normal star falls on to the neutron star, heating up to many millions of degrees as it does so. At such temperatures the gas emits X-rays. X-rays can be studied only by satellites in space because they do not penetrate Earth's atmosphere. Several satellites have been launched to study these bizarre objects.

Black Holes

But theory predicts that even more exotic objects, known as black holes, may be formed by a supernova explosion. If the remains of the burned-out star weigh at least three times as much as the Sun, the object's inward gravitational pull is so great that it won't stop at the stage of a neutron star. Nothing can prevent it from shrinking and shrinking ever further, getting progressively denser and denser until the gravitational field around it becomes so great that nothing can escape, not even light. The object has become invisible–a black hole.

Such an object may seem absurd, but the theory of gravity predicts that black holes should occur. The most astounding thing about them is that, at their centre, the star continues shrinking until it shrinks itself down to nothing. It has ceased to exist, although its intense gravitational field remains around it, like the smile of Lewis Carroll's Cheshire cat.

According to traditional theory, nothing can get out of a black hole. Things can fall in, however–including unwary space travellers. Black holes can suck in gas which, like the gas falling on to neutron stars, heats up and emits X-rays before it plunges forever into the hole. Therefore, black holes should be detectable by the X-rays they emit.

At least one likely black hole has already been found. It is an X-ray source known as Cygnus X-1, orbiting a visible giant blue star in the constellation of Cygnus. Gas falling from the blue giant star on to an invisible companion causes the X-ray emission. Calculations show that the invisible object on to which the gas is falling has a mass eight or 10 times that of the Sun. This is too heavy for a neutron star; therefore, it must be a black hole. There are other black-hole candidates, but none as definite as Cygnus X-1. The search for black holes in space with telescopes in orbit is one of the most exciting aspects of modern astronomy.

BEYOND THE FRINGE

What of the Milky Way as a whole? Only during the past 60 years or so have we obtained a reasonably accurate picture of where the Sun lies in the Galaxy, and how the Galaxy relates to the rest of the Universe. Before 1917 it was supposed that the Sun lay somewhere near the centre of a disk or lens-shaped aggregation of stars that formed the Milky Way. But in that year, the American astronomer Harlow Shapley (1885–1972) found that the Sun is nowhere near centrally placed in the Galaxy, but lies about two-thirds of the way to one edge. Therefore we are very much in the galactic suburbs.

Today, astronomers estimate there are at least 100 000 million stars in the Galaxy–the true figure may be two or three times as great. The entire Galaxy is so large that a beam of light moving at 300 000 km (186 000 miles) per second, takes approximately 100 000 years to cross it; therefore we say that the Galaxy has a diameter of 100 000 light years. By using optical and radio telescopes, astronomers have been able to trace the Galaxy's spiral shape. The Sun lies about 30 000 light years from the centre, in one of the spiral arms that curve out from the Galaxy's heart.

But even once our position in the Galaxy had been established by Shapley, it was still generally thought that nothing except space existed beyond the edges of the Milky Way. We now know how wrong that idea was. In 1923 the American astronomer Edwin Hubble (1889–1953) found that ours wasn't the only galaxy in the Universe. Using the 2.5-m (100-in) telescope on Mount Wilson in California, he photographed for the first time individual stars in the famous Andromeda nebula, a fuzzy patch of light just visible to the naked eye in the constellation of Andromeda. These stars were so faint compared with the stars close to us that he realized the Andromeda nebula must be far outside our own Galaxy. Nowadays, we know the Andromeda nebula is really a spiral galaxy of hundreds of thousands of millions of stars, similar to our own Milky Way, and that it

The spiral-shaped Andromeda galaxy is a twin of our own Milky Way, lying over 2 million light years away.

is 2.2 million light years distant. It is the most distant object visible to the naked eye, yet it is still one of the nearest galaxies to us.

Hubble went on to find ever more distant galaxies in space, classifying them into types according to their shapes–spiral, elliptical, or irregular. Each of these galaxies consists of a mass of stars. There must be thousands of millions of galaxies visible through the largest modern telescopes.

In 1929, Hubble made an astounding discovery which forms the basis of our modern understanding of the Universe. He found that all the galaxies seemed to be moving apart from each other, as though the Universe were expanding like a balloon being inflated. The most distant galaxies are moving away from us the fastest, so measuring a galaxy's speed allows us to estimate its distance.

The Big Bang

What could be the explanation of the expansion of the Universe? Georges Lemaitre (1894–1966), a Belgian astronomer, suggested that all the matter in the Universe had once been clustered together in super-dense form. Some explosive event, termed the Big Bang, set off the expansion, thereby marking the origin of the Universe as we know it. The galaxies are fragments from the explosion which have been flying apart ever since.

By measuring the rate at which the Universe is expanding, we can tell when the Big Bang must have happened. Latest estimates suggest that the Big Bang took place between 10 000 million and 20 000 million years ago. Therefore, modern astronomy allows us to date the creation of the Universe–assuming the Big Bang theory is right.

A modification of the Big Bang theory says that the Universe oscillates, alternately expanding and then contracting back to another Big Bang. On this theory, the current expansion of the Universe ought eventually to slow down and stop, but there is no evidence that this is happening. Observations of the motions of distant galaxies indicate that the Universe does not oscillate but

The American astronomer Edwin Hubble classified galaxies into ellipticals, spirals or barred spirals according to appearance.

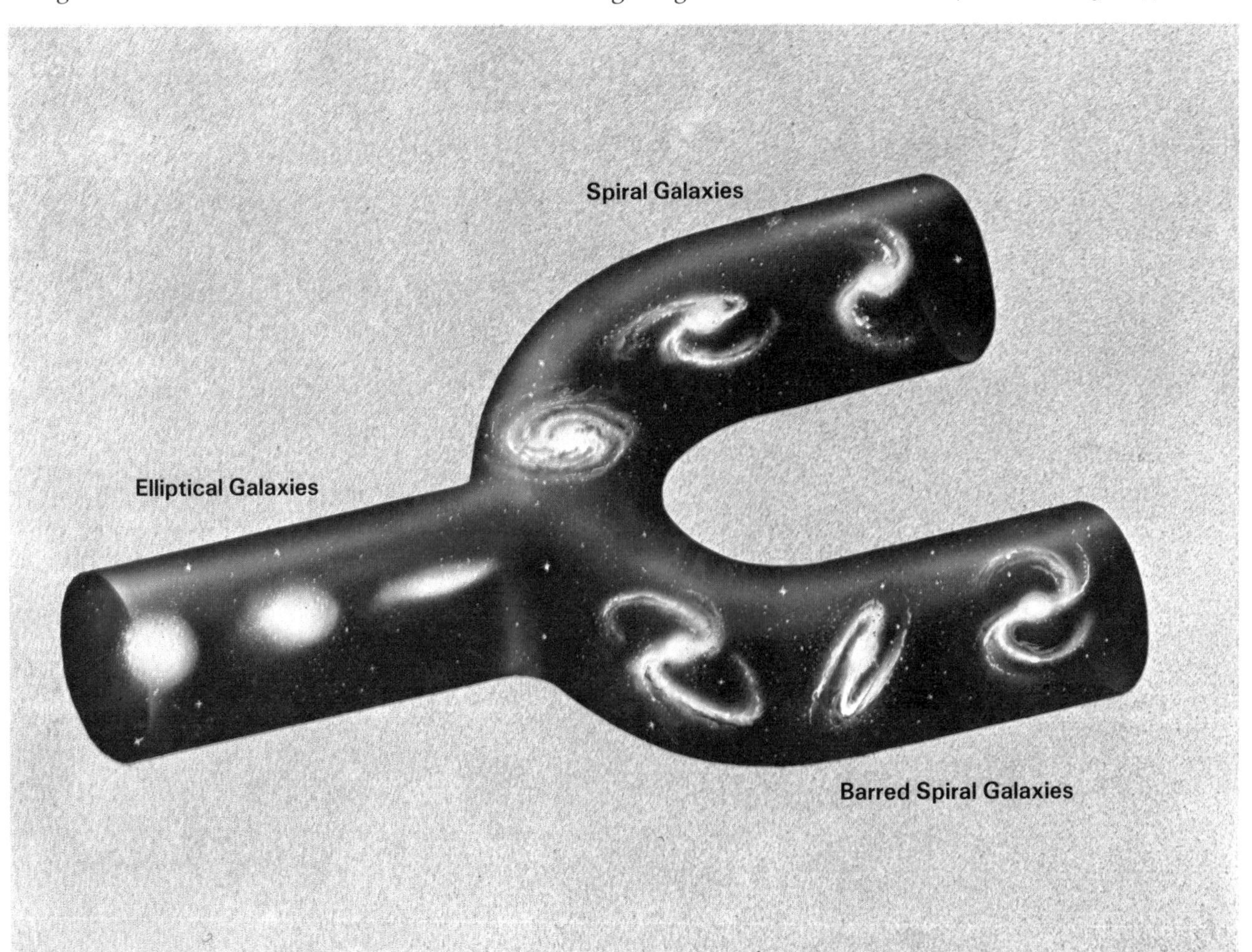

will continue to expand forever, slowly thinning out until the stars die and all becomes darkness.

One major rival to the Big Bang was the Steady State theory. This proposed that the Universe was not created all at once, but that new matter is very gradually coming into being in space. Therefore the Universe had no specific beginning and will have no end.

On this Steady State theory the Universe would look much the same at all times, unlike the Big Bang theory which says that the Universe has changed in appearance with time.

Quasars

But astronomers have found that the Steady State theory must be wrong because the Universe *did* look different in the past. They are able to discover this because, by looking deep into space, we are

According to modern views, our Universe began in a giant explosion called the Big Bang, from which it has been expanding ever since. In the Big Bang theory, as the Universe expands the galaxies in it thin out.

also looking back in time. Light from the most distant parts of the Universe takes a very long time to reach us. Far off in the Universe, as revealed by large telescopes, are strange objects called quasars. Many quasars are estimated to be at least 10 000 million light years away, judging from the speed at which they are moving in the general expansion of the Universe. Therefore their light has been travelling towards us for at least 10 000 million years, and so we see them as they appeared long before Earth was born, in the early days of the Universe.

Quasars are thought to be galaxies in the early stages of their evolution, giving out hundreds of times as much energy as our Milky Way. Most of a quasar's energy comes from a small area at its heart, and it is these brilliant centres that make quasars visible over such enormous distances.

Centaurus A is a strong radio source surrounding the visible galaxy NGC 5128.

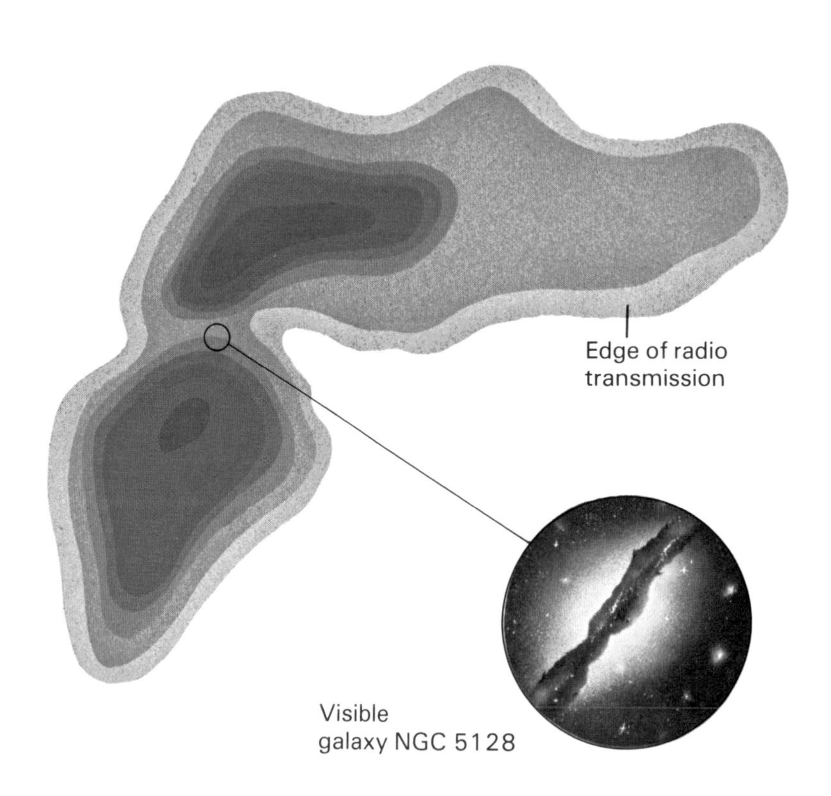

Telescopes launched into orbit by the Space Shuttle in the 1980s will allow astronomers to see even fainter and more distant objects.

As astronomers look deeper and deeper into space with optical and radio telescopes, they find ever more quasars and related objects called Seyfert galaxies and radio galaxies.

Seyfert galaxies are named after the American astronomer Carl Seyfert, who discovered them in 1943. They are spiral galaxies with brilliant centres, although less brilliant than quasars. Seyfert galaxies seem to provide the link between normal galaxies and quasars. Radio galaxies, as their name implies, are strong radio emitters.

Putting all these observations together, astronomers conclude that the Universe has been evolving from an origin between 10 000 million and 20 000 million years ago, as the Big Bang theory predicts. Further support for the Big Bang theory comes from the existence of a weak background radiation present throughout space. This was first discovered by radio astronomers in 1965, and is believed to be energy left over from the Big Bang explosion itself.

What is the energy source that powers quasars, Seyfert galaxies and the like? One of the most baffling facts about quasars and their relatives is that the central area from which most of their radiation comes is very small, often not much bigger than the solar system.

How can the radiation of hundreds of normal galaxies be generated in such a comparatively small space? And why should galaxies have been so much brighter and more energetic in the past than they are now?

No one knows for sure, but there are some intriguing theories. Chief among these is the suggestion that at the centres of quasars and galaxies lurk super-massive black holes, with the mass not of just a few suns but with millions of solar masses. These would begin with the supernova death of a few large stars, and grow by the influx of more material from the space around them, like a scaled-up version of the black hole in Cygnus X-1. But, unlike Cygnus X-1, they are sucking in not just gas but whole stars, and it is the chaos of star destruction around these super-massive black holes that makes quasar hearts so brilliant. Once all the stars in this central region have been swept up, the black hole's fuel source is removed, and the central brilliance dies down.

As would be expected with this theory, the centres of quasars and their relatives turn out to be emitters of X-rays. Further support for the theory comes from the discovery of a small 'hot spot' at the centre of our own Galaxy which has the expected size of a

An artist's impression of a quasar, thought to be a galaxy in the early stages of its evolution.

super-massive black hole. There may be greater similarities between quasars and 'normal' galaxies like our own than has previously been realized. The study by observatory satellites of the centres of distant galaxies and quasars, and of the centre of our own Galaxy, is an important and exciting new branch of space science.

Mini Black Holes

Astronomers are also interested in black holes of another kind–mini black holes. Such objects would, according to theory, have the mass of a mountain compressed into the size of an atomic particle–about a ten-million-millionth of a centimetre. Mini black holes could not form in the Universe today, but the English physicist Stephen Hawking (born in 1942) has shown that they could have been produced in the highly turbulent conditions of extreme density and pressure which prevailed shortly after the Big Bang. Find a mini black hole, therefore, and you find direct evidence of the Big Bang origin of the Universe. But how to detect such a microscopic object?

Hawking has shown that something remarkable happens to mini black holes, which would give away their existence. It turns out that the gravitational field around a black hole is not as impenetrable as has previously been thought, but instead is slightly leaky. This leakiness is greatest for the smallest black holes, which can radiate energy like a hot body and can even explode! This research by Hawking is at the forefront of modern physics, and has major implications for science in general.

In practice there is hardly any leakage from large black holes like that in Cygnus X-1 or at galaxy centres. But the leakage effect becomes important for the mini black holes predicted to have been formed in the Big Bang. Some of these, if they exist, should be exploding today, emitting a burst of gamma-ray radiation which could be picked up by large detectors placed in orbit. The Space Shuttle may be used to carry equipment for detecting the explosion of mini black holes, the discovery of which would provide important new information about the origin of the Universe, as well as confirming a remarkable prediction of modern physics.

Our survey of the Universe has brought us a long way from the solar system, where we started. We have ranged from the nearest planets to the most distant galaxies and quasars, and have even glimpsed the possible way in which the Universe was created. Space research has helped us reach our modern understanding of the Universe and is still helping us unravel the problems that remain.

Let us now concentrate on the ways in which mankind has attempted to explore the space around us, by rockets, satellites and space probes.

The Space Shuttle, America's re-usable space plane of the 1980s will be able to carry equipment that may detect the existence of mini black holes.

2 THE SPACE AGE BEGINS

No one knows exactly when and where the rocket was invented; the truth is lost in the mists of time. But the invention is usually credited to the Chinese, who were seemingly employing rocket-propelled arrows by the thirteenth century, and possibly two centuries prior to that. These were firework-type rockets, propelled by burning a black powder (sulphur, saltpetre and charcoal) commonly termed gunpowder.

Powder-propelled rockets were used sporadically in warfare until the modern jet and missile age, when rockets took on an alarmingly new and powerful form. Over all those centuries, rocketeers were making use of the basic principle that would put men in space.

By the thirteenth century, the Chinese used rocket-propelled arrows in warfare.

HOW A ROCKET WORKS

The principle governing all rockets, from fireworks to giant space launchers, is that of reaction to a force. In a rocket, the force is caused by the high-speed escape of hot gases out the back; the energy of the escaping gases pushes against the rocket body, driving the rocket forward. This principle is embodied in Newton's third law: to every action there is an equal and opposite reaction. Therefore, the stronger the force of the exhaust, the stronger the reaction on the rocket.

For a simple demonstration of the rocket principle, blow up a balloon. The air inside the inflated balloon is the exhaust gas, although in a real rocket the exhaust gases are produced by burning fuel. Let the balloon go and air streams out of the nozzle, driving the balloon forward. A balloon careers crazily around the room because it is not rigid, but a space rocket is built so that the exhaust gases escape in a controlled manner; the exhaust nozzles can be swivelled slightly from side to side to steer the rocket.

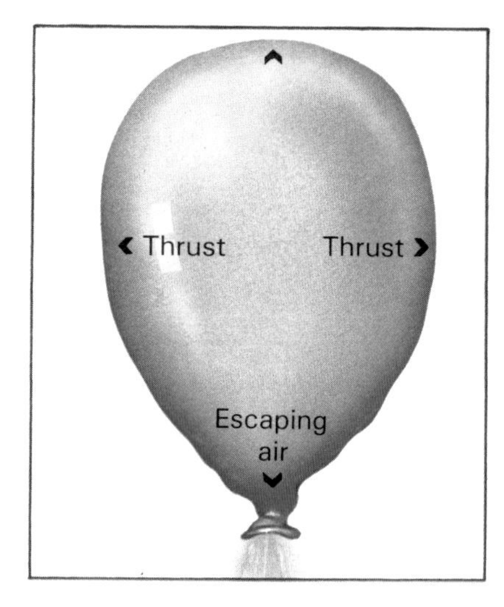

Principle of the rocket. A rocket is propelled by the thrust of escaping gases, as is an inflated balloon that is let go. In a balloon, the thrust is produced by air on the inner surface of the balloon. In a rocket, the thrust is produced on the inner surface of the combustion chamber and nozzle.

To get into space, a rocket must travel fast enough to overcome the effect of Earth's gravity, which tries to pull it back. In practice, this means reaching a speed of 8 km (5 miles) per second to get into orbit around Earth, or 11.5 km (7 miles) per second to escape from Earth entirely and reach the Moon or another planet.

Imagine throwing a ball upwards. It first rises and then falls back under the pull of gravity.

Throw it harder, and it rises further before falling back. If you could throw the ball at 8 km (5 miles) per second, it would travel so far that it would go into orbit around Earth, as long as you threw it at the correct angle to the ground.

This is the job that a space rocket does. Although at first it climbs vertically, the rocket soon begins to tilt over, under the control of its on-board guidance system, so that as it reaches its top speed it is moving parallel to the ground and it slides into orbit around Earth, as shown in the diagram on the right.

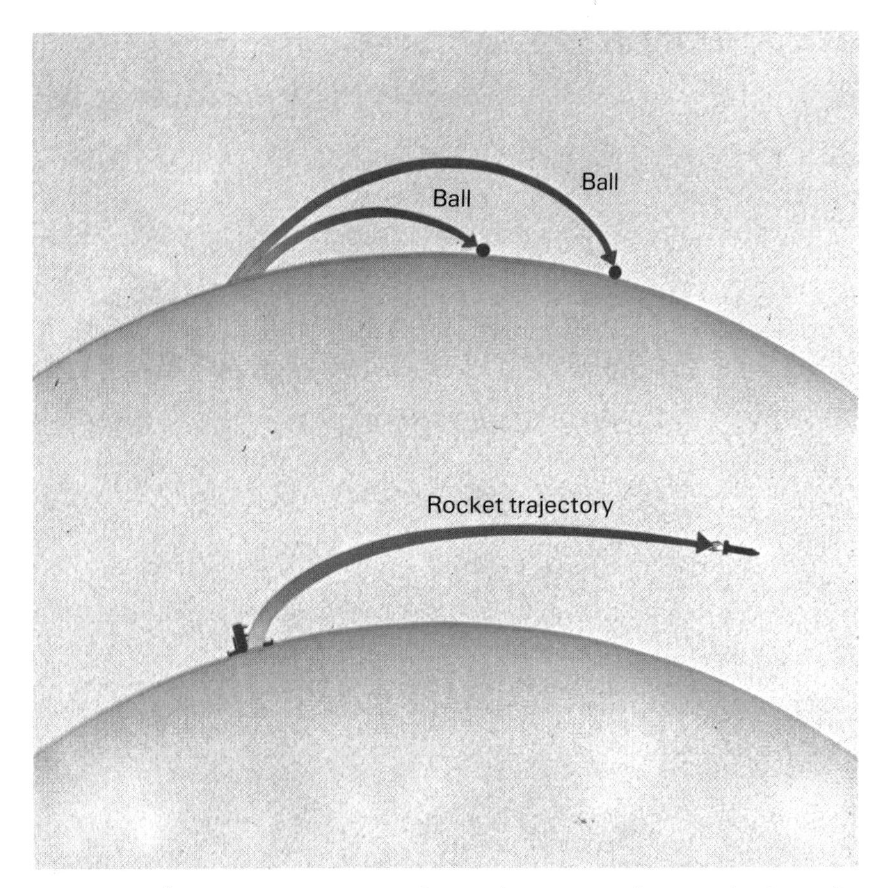

Right: Getting into orbit. A ball thrown into the air rises before falling back again under the pull of Earth's gravity. Thrown harder, the ball travels further before falling back. If the ball were thrown at 8 km (5 miles) per second on a suitable trajectory, it would go into orbit around Earth. A rocket can provide the necessary power to launch an object into orbit.
Below: A simple rocket. The fuel and oxidizer are held in separate tanks. They are pumped into the combustion chamber and ignited to produce the hot gases that propel the rocket.

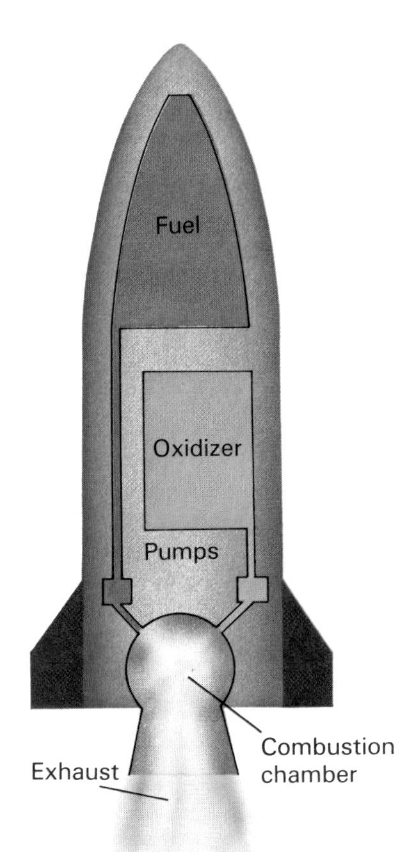

In practice, a great deal of complicated equipment is necessary to ensure that the job is properly done. Most modern rockets use liquid propellants; typical fuels are kerosene or liquid hydrogen. Some space rockets still use advanced versions of the time-honoured black powder; these are known as solid propellants.

In a liquid-propellant rocket two tanks are required, one to hold the fuel itself while the other contains an oxidizer, which is a substance that supports combustion of the fuel even in the airless vacuum of space. In air it is oxygen that supports burning, so rockets usually employ liquid oxygen as the oxidizer.

Pumps drive the fuel and liquid oxygen from their tanks into the combustion chamber, where they are ignited by a spark and burned to produce hot gases. These gases escape at high speed through the exhaust nozzle, producing the thrust that drives the rocket forwards. The faster the exhaust speed, the better; space rockets have exhaust speeds approximately ten times the speed of sound and their launch is accompanied by sonic booms that shake the ground like loud thunderclaps.

Because Earth's gravity is so strong, space rockets have to carry a lot of fuel to enable them to break free, which makes them big and heavy. For instance, the Saturn V Moon launcher, which stood 111 m (364 ft) tall, carried nearly 3000 tonnes of fuel; its empty weight was 200 tonnes, in addition to the actual payload.

The structural weight of tanks needed to carry all that fuel, and the giant engines required to get the rocket off the ground, become unnecessary when the rocket has taken off and the fuel tanks are becoming empty. This weight prevents the rocket from reaching the necessary orbital speed of 8 km (5 miles) per second.

The Multi-Stage Rocket

The solution is the multi-stage or step rocket, in which the unwanted dead weight of fuel tanks and engines is discarded as the rocket ascends. By adding another smaller rocket on top of the first one, so that it starts burning once the first has exhausted its fuel and dropped away, the final payload will fly further and faster than that of a single large rocket. The first stage acts as a springboard into space for the upper stage or stages.

Most space rockets are built in two or three stages, though some launchers can have up to five stages. When the stages are stacked on top of each other it is termed series staging. But extra stages can also be added at the side of the main rocket; this is called parallel staging. Such parallel stages are often referred to as strap-on boosters. They fall away as the main rocket ascends. Large Soviet rockets use this system, and strap-on boosters are also employed by the United States for its Titan III and Space Shuttle.

Rockets are usually held down by clamps to ensure that all engines are burning correctly before launch. Then the rocket is released and climbs slowly into the sky. Were a rocket to accelerate too quickly, the payload–men or instruments–could be damaged by the forces involved. As the rocket gets above the dense layers of Earth's lower atmosphere, its acceleration increases. In fact, rockets work best in the vacuum of space where there is no air resistance.

Were Earth smaller, its gravity

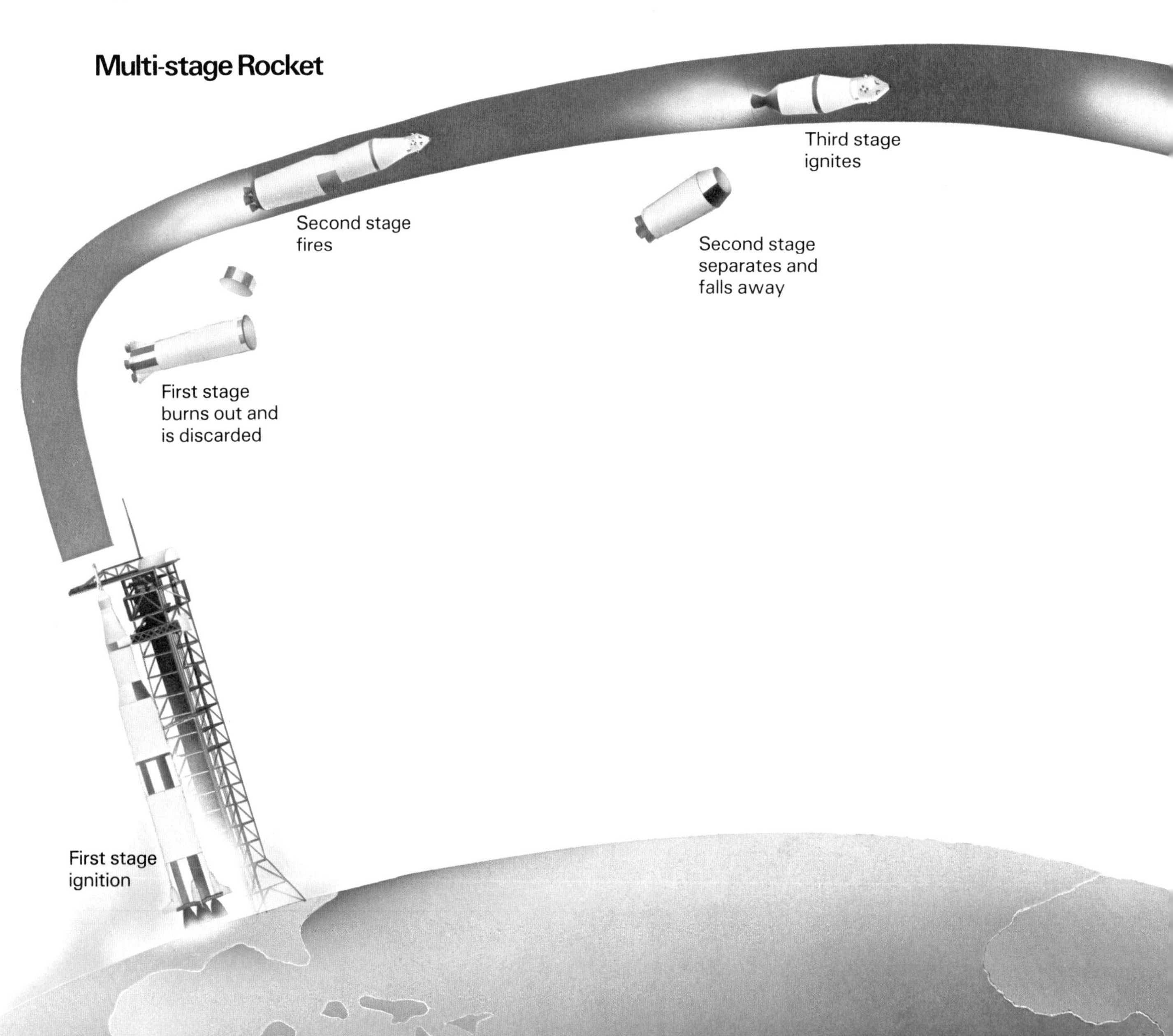

would be less and rockets could be smaller. For instance, the Moon has only one-sixth of Earth's gravity and so two astronauts were able to take off from its surface in the small craft called the lunar module. A spacecraft can orbit an airless body like the Moon at any height, because there is no atmosphere to resist its motion. But, because of the dense, deep atmosphere of Earth, satellites have to be put into orbits at least 150 km (93 miles) or so above our planet's surface. Even at these heights, however, there is still some thinly-spread gas from the outer atmosphere, which resists the motions of satellites and gradually slows them down until they fall back to Earth. They burn up by friction as they plummet into the dense layers of the atmosphere, causing a brilliant fireball visible to observers on the ground. Very occasionally, a lump of metal gets through the atmosphere to crash on Earth, as happened with parts of the Russian Cosmos 954 satellite which fell in north-western Canada in January 1978.

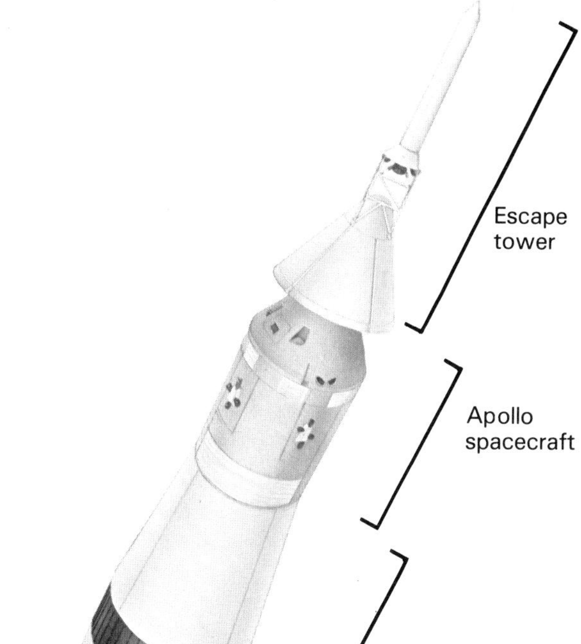

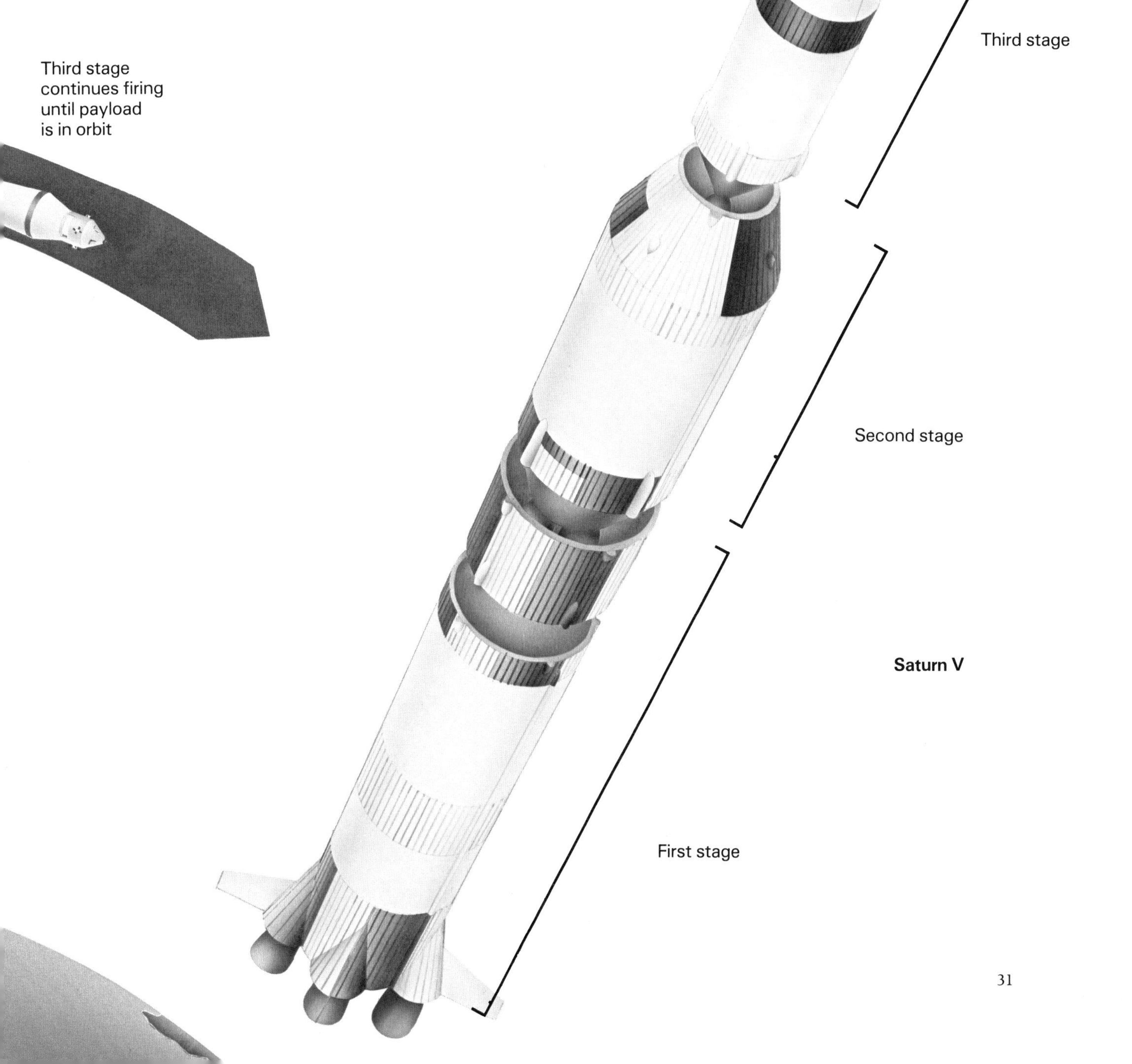

HISTORY OF THE ROCKET

Only since 1957, when the first Sputnik was launched, have rockets been powerful enough to put satellites into orbit around Earth. To the early rocketeers with their black powder rockets, space travel, if they ever thought about it, was an impossible dream. Their rockets were intended for warfare. At the capture of Baghdad in 1258 the Mongols used rockets which the Arabs called 'Arrows from Cathay'. Indian soldiers used rockets with such effect against the British at the battles of Seringapatam in 1792 and 1799 that the British soldier William Congreve (1772–1828) decided to adopt them for his own forces.

He designed a new and improved form of incendiary rocket weighing 14.5 kg (32 lbs) with a range up to 3000 m (9840 ft) for the successful attacks against Boulogne in 1806 and Copenhagen in 1807. Congreve rockets were used during offensives in the war of 1812 against America; they are referred to by the phrase 'The rocket's red glare' in the American national anthem. These rockets had considerably improved range and accuracy, but still needed cumbersome long sticks to stabilize them.

Another Englishman, William Hale (1797–1870) improved rockets still further by drilling angled vents around their base so that they spun in flight; this spin stabilization did away with the need for sticks. Hale later introduced curved vanes into the rocket's exhaust stream to effect the spin stabilization, and his designs superseded those of Congreve by 1867. Subsequently, though, the use of rockets for warfare declined in the face of improved artillery–that is, until the advent of the V2 in the Second World War.

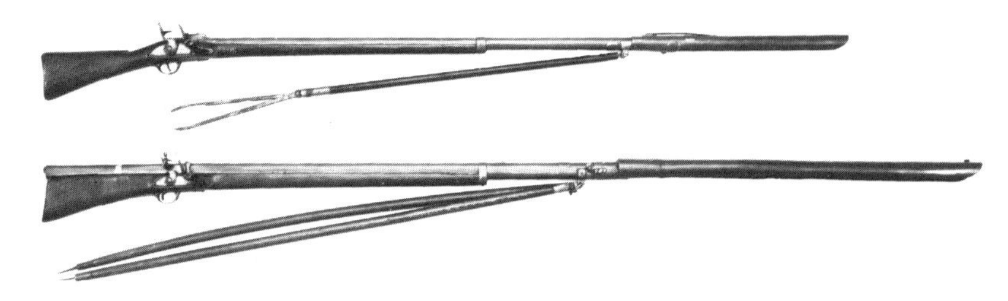

Flintlock firearms, with supporting legs, used for firing small rockets c. 1810 to 1820.

Science Fiction–or Fact. . . .?

Many people dreamed of space travel long before it ever became possible. The Greek writer Lucian of Samosata in the second century AD wrote what is the first known work of space fiction, misleadingly titled *True History*. In it, he fancifully imagined that a sailing vessel is lifted into the air by a whirlwind, which bears it to the Moon. There, the sailors encounter lunar inhabitants called the Hippogypi, who ride on three-headed vultures. Lucian's follow-up to this tall tale concerned a hero who flew from Mount Olympus to the stars using birds' wings.

Lucian's stories are characterized more by imagination than scientific plausibility for there is, of course, no air in space and wings would thus be useless, although this fact was not known at the time. The German mathema-

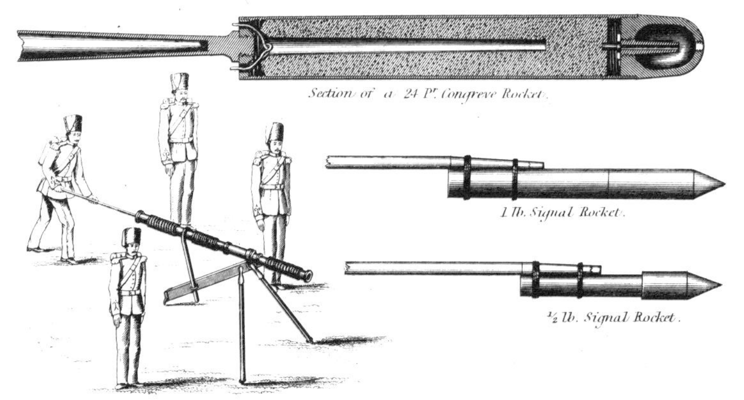

The British soldier William Congreve introduced rockets like these, *above left*, after suffering heavy losses to Indian rocket forces at the battle of Seringapatam, *left*.

tician Johannes Kepler (1571–1630), best known for his laws of planetary motion, wrote an amazingly imaginative work called *Somnium* ('Dream'), in which demons journey to the Moon by supernatural means; Kepler knew that ordinary flight would not work in the vacuum of space. His main aim seems to have been to describe astronomical knowledge of the Moon, but he chose also to populate the surface with strange cave-dwelling beings.

Bird's-eye View

Wing-powered flight to the Moon was revived by Francis Godwin, Bishop of Hereford, who wrote a story called *The Man in the Moone,* published posthumously in 1638 under the pseudonym of Domingo Gonsales. In the story, Domingo Gonsales and his servant, marooned on an island, trained wild geese to carry them to safety. The geese turned out to be so efficient that they carried Gonsales to the Moon, taking 11 days for the trip. There he found the Moon to be like another Earth, largely covered by water and populated by large and colourful creatures.

Cyrano de Bergerac published a science fantasy called *Voyage to the Moon* in 1649. He attempted to make his journey by attaching to himself flasks of dew; as the dew evaporated in the heat of the Sun, he was carried upwards. He failed to complete the trip with this device, crash-landing on Earth instead. His second attempt, using a rocket-powered device, proved more successful.

Even Daniel Defoe, of *Robinson*

Below left: Royal Naval Rocket Brigade in Abyssinia, 1868.
Below: Rocket boats bombarding Sveaborg during the Crimean War, 1855.
Bottom: Rocket practice on Woolwich marshes, 1845.

Top left: Jules Verne. These illustrations from his book show:
Top right: The space projectile.
Above left: The cannon which fired it.
Above right: Weightlessness.
Centre: Inside the padded capsule.

Crusoe fame, wrote a tale of lunar travel called *The Consolidator* in 1705; the consolidator itself was a spacecraft with wings.

In 1752 the French author Voltaire produced his famous satire called *Micromégas,* which is the name he gave to a gigantic hero from the star Sirius. Micromégas travels the Universe, visiting our own solar system for a sojourn with the supposed inhabitants of Saturn. Eventually, Micromégas reaches Earth and is surprised to find it populated with what, to him, are tiny beings. This book was an attempt to put into perspective man's insignificance in the cosmos.

To the Moon in a Balloon!

Edgar Allan Poe in 1835 sent his hero Hans Pfaall on a lunar trip in a home-made balloon, equipped with apparatus to condense the rarefied air he believed existed in space so that Pfaall could breathe. Pfaall's trip was made to escape his debts on Earth. After 17 days, the balloon burst and he crashed on the Moon, where he found himself in a lunar city. There he remained, becoming history's first celestial tax exile.

Most celebrated of all space fiction works are those of Jules Verne. In his now-classic 1865 book *From the Earth to the Moon,* Verne imagined a projectile containing three men and two dogs, fired towards the Moon by a 300 m (984 ft) cannon. In reality, the tremendous acceleration forces in a gun would squash the space travellers flat before they left the barrel, and air resistance at those speeds would have burned up the projectile, so Verne's idea is not as practicable as it must have sounded to his readers. But Verne had two uncanny strokes of prescience; he placed the launching site in Florida, USA, not far from the present-day Cape Canaveral, and he had the whole

Top: H. G. Wells, and an illustration from *The War of the Worlds.*
Above: Illustration for Arthur C. Clarke's *The Exploration of the Moon,* 1954.

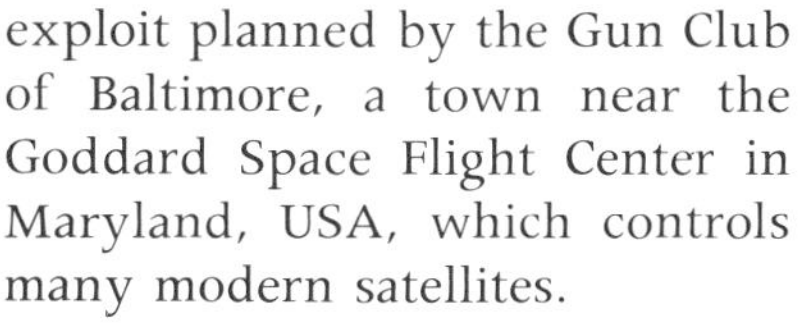

exploit planned by the Gun Club of Baltimore, a town near the Goddard Space Flight Center in Maryland, USA, which controls many modern satellites.

The Shape of Things to Come
In Verne's spectacular story, which was concluded in the sequel *Around the Moon,* the space voyagers had their course perturbed by the close passage of an asteroid. This put them into orbit around the Moon rather than landing on it as originally planned. By firing what we would today call retro-rockets, the intrepid travellers were able to return to Earth where they splashed down in the Pacific—another foretaste of the real Apollo missions.

The first proposal for an artificial satellite around Earth seems to have come in 1869 from the American author Edward Everett Hale. In his story *The Brick Moon,* Hale imagined an artificial satellite made of brick to be launched into a polar orbit to help sailors with navigation. The brick sphere, 61 m (200 ft) in diameter, was to be launched by rolling against a fast-spinning flywheel; but it slips too soon and is hurled aloft, along with several construction workers and their visiting families, who continue orbiting happily in their space station.

H.G. Wells wrote two classics of space fiction, *The War of the Worlds* (1897) and four years later *The First Men in the Moon. The War of the Worlds* deals with a frightening invasion of Earth by Martians, whereas *The First Men in the Moon* reviews the theme of lunar travel—in this case, a spaceship powered by an anti-gravity device. The two lunarnauts find the Moon covered with vegetation and inhabited by a subterranean race of Selenites. Even though Wells' story flew in the face of scientific knowledge, it was still highly successful.

TSIOLKOVSKY, THE TRUE SPACEFLIGHT PROPHET

By the time of Verne and Wells, space fact was starting to overtake space fiction, although few people then realized it. An obscure Russian schoolmaster, Konstantin Eduardovich Tsiolkovsky (1857–1935), was beginning the theoretical work that would make him the first true prophet of spaceflight. Tsiolkovsky had become partially deaf at the age of 10, which added to his already introverted and studious nature. He was fascinated by the idea of flight and began experiments in aerodynamics.

Long before the Wright brothers had made their first aircraft flights, Tsiolkovsky's speculations had soared far beyond the atmosphere. In 1883 he realized that the action-reaction principle of rockets made them ideal for space travel, an idea which seems obvious to us today but which had been missed by almost everyone up until then.

Tsiolkovsky said the idea was inspired by the novels of Jules Verne, but Tsiolkovsky went much further than Verne by working out mathematically the theoretical principles of rocketry. In 1898 he hit upon the formula that governs rocket propulsion, showing how the rocket's final speed is related to its exhaust velocity and how much propellant is needed for a rocket of given mass.

Tsiolkovsky's writings described the conditions of weightlessness experienced in space, and also contain the proposition for an artificial satellite.

In 1903 he conceived the idea of a liquid-fuel rocket using liquid hydrogen and liquid oxygen. Subsequently, he worked out the mathematical principles of the multi-stage rocket, which he termed a rocket-train, showing that such an arrangement was virtually essential to escape from Earth. Tsiolkovsky singlehandedly discovered the basic principles of space travel, although only at the end of his life did he begin to receive the recognition he was due.

Below: Model of a rocket designed by Konstantin Tsiolkovsky at a museum in his home town of Kaluga.
Right: Monument to Tsiolkovsky in Kaluga.

Below: Konstantin Tsiolkovsky, Russian prophet of space travel.
Right: Tsiolkovsky, on right of picture, in his workshop with writer K. Altaisky.

GODDARD'S ROCKETS

Tsiolkovsky was solely a theoretician, not an experimenter, of rocketry. The credit for many practical developments in rocketry goes to an American physicist, Robert Hutchings Goddard (1882-1945), who built and flew the world's first liquid-fuelled rocket.

Goddard's interest in spaceflight was inspired by the science fiction of Wells and Verne. He became professor of physics at Clark University in Worcester, Massachusetts, USA, where he took out his first patents on rocket propulsion systems and multistage rockets. Throughout his life he registered over 200 patents, covering many aspects of rocket propulsion; in 1960 the US Government belatedly paid one million dollars for use of these patents to his widow and to the Guggenheim Foundation, which supported many of his experiments.

Goddard realized that liquid-fuelled rockets were the best for spaceflight, and he intended to car y out experiments with them. To do so he needed money and so he submitted a paper to the Smithsonian Institution of Washington which described how a rocket could be used for scientific purposes such as upper-atmosphere research. This brought him his first modest grant.

Goddard's paper for the Smithsonian, published under the title *A Method of Reaching Extreme Altitudes,* was mostly devoted to outlining the basic principles of rocketry. But it also contained a small section in which Goddard speculated on the possibility of sending a rocket to hit the Moon, setting off a mass of flash

Robert H. Goddard, the American rocket pioneer, seen in his workshop at Roswell, New Mexico, in 1935 with a rocket under development.

powder. To Goddard's surprise, this sensational idea was widely publicized by the Press, which subjected him to much hurtful and unwarranted criticism.

Blast Off!

By 1926 Goddard had progressed sufficiently to build a primitive but practical liquid-propellant rocket which burned gasoline and liquid oxygen. The world's first liquid-fuelled rocket flew on March 16, 1926. The rocket was airborne for a mere two and a half seconds; in that time it reached an altitude of 12.5 m (41 ft) at an average velocity of 100 kph (62 mph). But it was a start.

Goddard continued his experiments at Worcester, Massachusetts, until 1929. By then, he had made four flights of liquid-propellant rockets, and attracted more unwanted publicity. Charles Lindbergh, the famous aviator, became interested in Goddard's work, and arranged a substantial grant for him from the Daniel Guggenheim Fund. With this backing, Goddard moved to a ranch near Roswell, New Mexico.

There, he continued to develop rockets powered by gasoline and liquid oxygen, helped by only four assistants. These improved rockets embodied new stabilization and combustion techniques, enabling them to reach near-supersonic speeds and climb up to 2 km (1.2 miles) in altitude.

Seeking ever more ways to improve his rockets, Goddard next introduced propellant pumps to force the liquids into the combustion chamber, launching rockets of this type up to 6.6 m (21 ft) in length and weighing 200 kg (440 lbs) or more. When his experiments were finally interrupted by the Second World War, Goddard had developed the basic components of a successful long-range rocket.

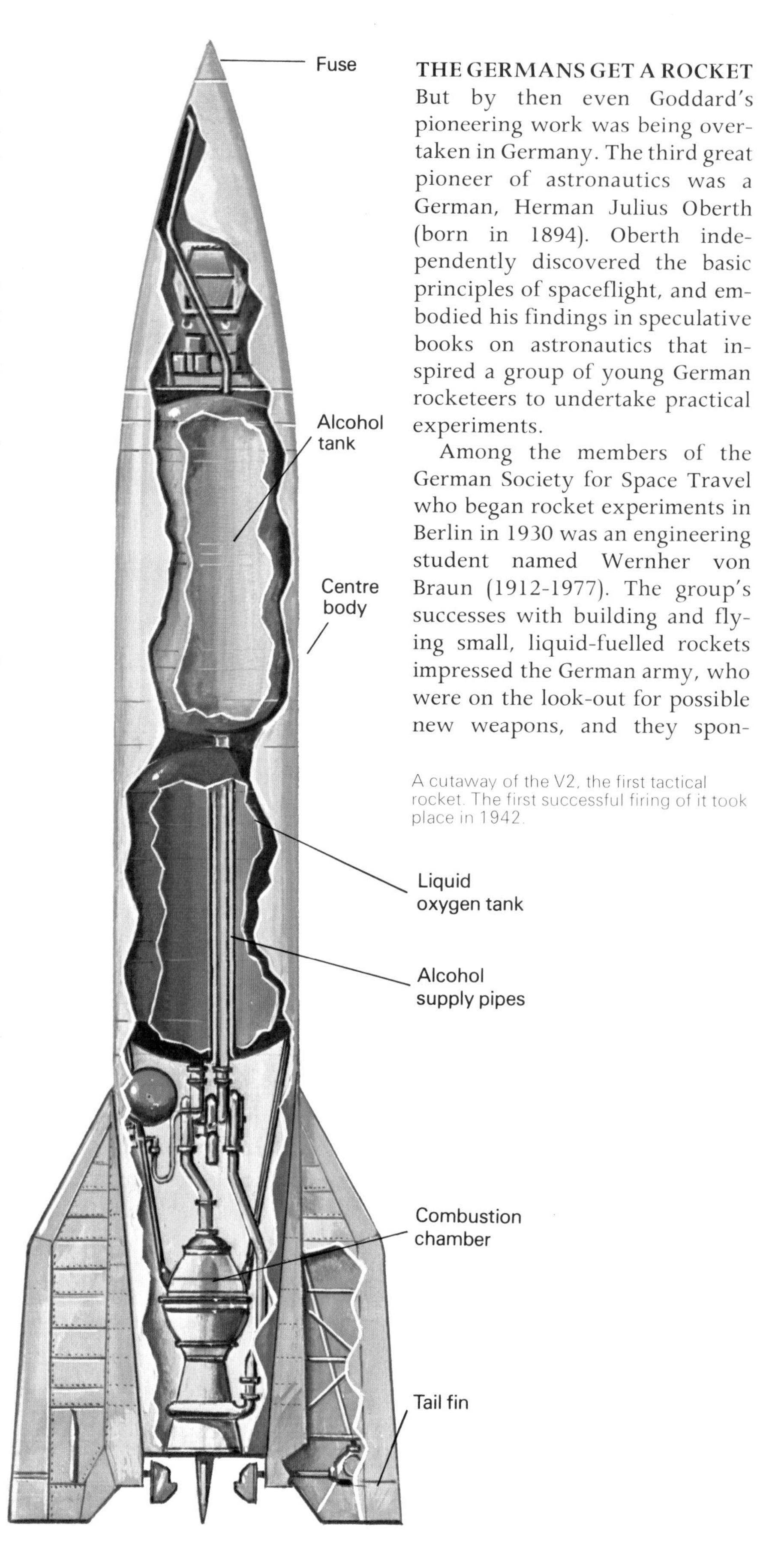

A cutaway of the V2, the first tactical rocket. The first successful firing of it took place in 1942.

THE GERMANS GET A ROCKET

But by then even Goddard's pioneering work was being overtaken in Germany. The third great pioneer of astronautics was a German, Herman Julius Oberth (born in 1894). Oberth independently discovered the basic principles of spaceflight, and embodied his findings in speculative books on astronautics that inspired a group of young German rocketeers to undertake practical experiments.

Among the members of the German Society for Space Travel who began rocket experiments in Berlin in 1930 was an engineering student named Wernher von Braun (1912-1977). The group's successes with building and flying small, liquid-fuelled rockets impressed the German army, who were on the look-out for possible new weapons, and they sponsored von Braun's research on rocket combustion. Von Braun joined Captain Walter Dornberger near Berlin, where development work began on the forerunners of what was to become the V2, a rocket fired against Britain at the end of the War.

In 1937, the Dornberger/von Braun rocket group moved to Peenemünde on the Baltic coast where, under top-secret conditions, they began work on the V2 proper. It was astoundingly large for its time: 14 m (46 ft) high, weighing 12.5 tonnes and capable of carrying a one-tonne payload up to 320 km (around 200 miles). It burned alcohol and liquid oxygen.

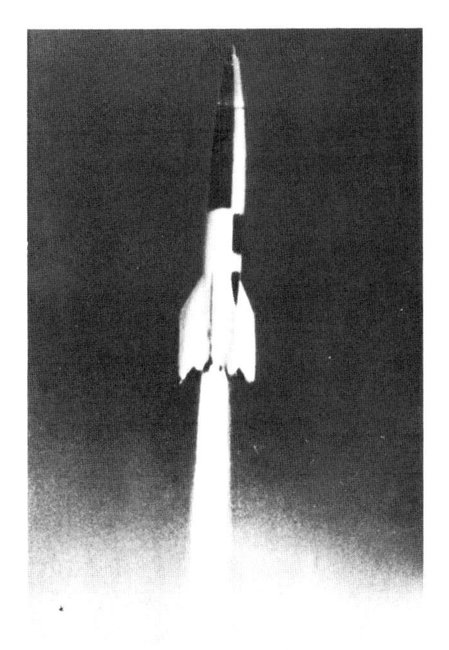

Left: Test flight of a V2 at Peenemünde in 1943.
Top: V2 assembly plant captured by American troops.
Above: A V2 being prepared for firing from a mobile launcher.

The first successful firing of a V2 took place on October 3, 1942, when the rocket covered a range of 190 km (118 miles) and reached a maximum altitude of 85 km (53 miles) and, incidentally, became the first rocket to exceed the speed of sound. Production of the V2 then got under way in earnest at a facility in the Harz mountains where over 5000 were manufactured. They came into service in September 1944 but were too late to alter the outcome of the War. Nevertheless, the V2 was to influence thinking about future wars.

To his credit, von Braun's ultimate aims lay far beyond the

bombardment of Earth. Space travel had always beckoned as the real goal of rocketry; and at Peenemünde he and his associates drew up designs for several advanced rockets, including a three-stage vehicle capable of putting man into orbit. Von Braun would eventually carry through his dreams, but it was to take a long while.

At the end of the War, von Braun and his team surrendered to the advancing American forces and were transferred to the White Sands Proving Ground in New Mexico. Here they continued development of the V2, this time for upper atmosphere research.

In 1950, von Braun moved to the army's Redstone Arsenal in Huntsville, Alabama, to work on missile programmes. Out of this came the rocket known as the Redstone, a medium-range military missile that was to play an early role in the US space programme.

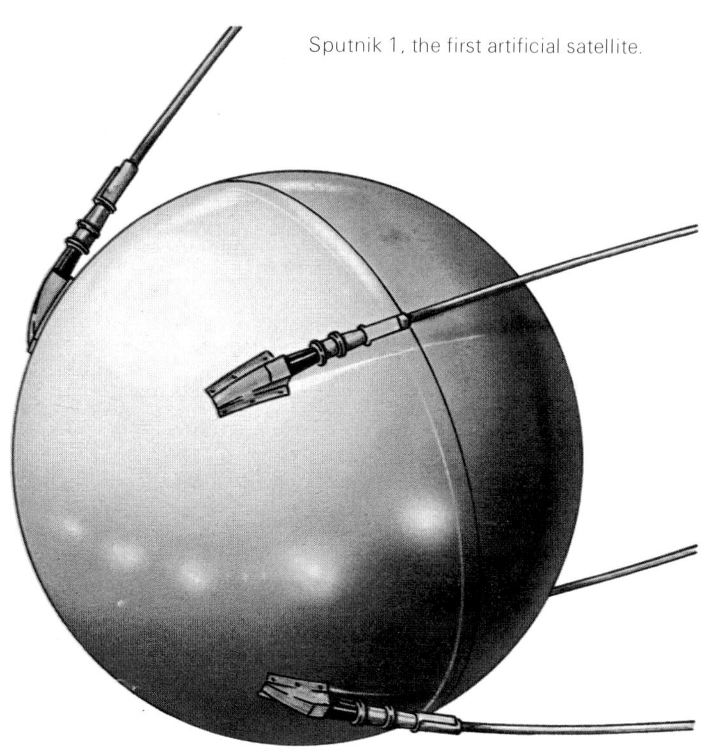

Sputnik 1, the first artificial satellite.

SOVIET EXPERIMENTS

Meanwhile, in the Soviet Union the legacy of Konstantin Tsiolkovsky had not been forgotten. During the 1920s a Russian engineer, Fridrikh A. Tsander (1887–1933), worked on designs for liquid-propellant engines, which he built and flew shortly before his premature death. Other Soviet rocket experimenters of the time included Valentin P. Glushko (born in 1906), Sergei P. Korolyov (1907–1966), and Mikhail K. Tikhonravov (1900–1974), all of whom became leading figures in the development of Soviet ballistic missiles after 1945.

Since most of the German rocketeers from Peenemünde joined the American military effort, the Soviet Union had to rely on its own ingenuity for missile design. Experiments were carried out with captured V2 rockets after the War, but the Russians had bigger ideas, particularly after their development of a hydrogen bomb in 1953. Under Korolyov's guidance as chief designer, a massive booster was created to carry a nuclear warhead around the globe. The rocket had to be big because the bomb was big. By contrast, the United States miniaturized its bombs, thus requiring smaller missiles.

To lob their heavy payload into space, the Soviet designers chose to use parallel staging–in this case, four boosters strapped around the rocket's central core. All five sets of engines ignited at lift-off, and the four strap-on boosters fell away as the vehicle ascended. Technically, it may not have been a particularly elegant or efficient scheme, but it worked. The first Soviet intercontinental ballistic missile flew successfully in August 1957, 15 months before its smaller American counterparts.

Such a rocket could also be used to launch satellites into orbit around Earth. This would be a tremendous propaganda coup, as the Soviets realized. The United States was openly planning an Earth satellite as part of the International Geophysical Year, a world-wide co-operative programme between Earth scientists. The Soviet Union announced that it, too, would launch a satellite, but no one took them seriously–until too late.

Sputnik I

On October 4, 1957, the space age began in earnest when Sputnik 1, the world's first artificial satellite, was orbited by a modified version of the Soviet intercontinental ballistic missile. To emphasize their space superiority, a month later the Soviets put into orbit a capsule containing a dog, while the US space programme fizzled on the ground.

Never before had East beaten West so comprehensively and so spectacularly. It created a deep-seated sense of rivalry that was not fully vanquished until the first American set foot on the Moon 12 years later.

THE US STEPS IN

Two main long range missiles were developed by the US Air Force. First was the Atlas, a remarkable missile with a very thin skin to save weight. At launch, a central engine and two outer engines fired. As the rocket ascended, the two outer engines fell away, along with a skirt around the lower part of the rocket, leaving the central engine to thrust the vehicle into space. Because of this arrangement, Atlas is often termed a one-and-a-half-stage rocket. The Atlas subsequently played a major role in modified form as a launcher in the US space programme, both on its own and with upper stages. Titan was a bigger and more powerful two-stage missile, also subsequently adapted for satellite and manned launches.

Above: Minuteman III is a solid-fuelled missile.
Below: Firing of a US Air Force Atlas missile. Atlas rockets have been used for many space launches.
Below right: The US Air Force Titan missile was a two-stage rocket modified for space use.

Largely for political reasons, the United States at first chose not to use adapted military missiles for its civilian space programme, but to develop a new launcher without defence connections instead.

The launcher, called Vanguard, was a three-stage affair based on a successful atmospheric sounding rocket called Viking. But in the panic that followed the launch of Sputnik, attention turned to von Braun's group at Huntsville. Here a modified version of the Redstone medium-range missile had already been developed that could orbit a satellite. Previous requests by von Braun's group to launch a satellite had been turned down, but suddenly the decision changed. The four-stage launcher, called Juno 1, placed the first US satellite into orbit from Cape Canaveral, Florida, at the end of January 1958. Vanguard, after two well-publicized failures, eventually launched the second US satellite in March, 1958.

Ever-Expanding Space Programmes

After Earth satellites came lunar and planetary probes, and manned spaceflights. The Soviet

Union added an upper stage to its standard launcher, thereby gaining sufficient power to send its first Luna probes to the Moon; in April 1961 this same rocket design put the first man, Yuri Gagarin, into orbit, in a Vostok capsule. Unlike the United States, whose space programme was carried out in the open, the Russians kept their space activities cloaked in secrecy. Not until 1967 did they publicly unveil their standard launch rocket, allowing everyone to see the vehicle which had ushered in the space age. Even today, only a limited amount of information is available about the Soviet space programme, despite the immense interest that it has created.

By adding a longer and more powerful top stage to their standard launcher, the Soviet Union created a rocket that could send probes to the planets, as well as orbiting heavier manned vehicles, such as the multi-man Voskhod and Soyuz series (see Chapter 4). This rocket, that started out as a bomb-carrying missile, has had a glorious career as a space launcher; it remains the only Soviet booster considered reliable enough to launch men into space. A much larger and more powerful Soviet rocket does exist, called Proton, which is used to launch current Soviet lunar and planetary probes, as well as the Salyut space station. But Proton rockets have suffered spectacular launch failures, and they have never yet been risked for a manned launch.

In the United States' space programme, a baffling array of launchers has been used, based on the wide range of missiles developed for military purposes. Among the most successful of these was a launcher based on the Thor medium-range missile; over the years this was developed into the Delta rocket, the workhorse of the US space programme until the introduction of the re-usable Shuttle.

We have already mentioned the US Atlas missile which was pressed into service as a space launcher, first with an upper stage called Agena and later with the more powerful Centaur. Atlas rockets were used to boost astronauts into orbit in the Mercury series. For the subsequent two-man series of Gemini space launches the Titan missile was used. Titan rockets in various guises have been used for satellite launches. Improved versions of the Titan have two large strap-on boosters attached.

Among the spacecraft launched by up-graded Titan rockets are military reconnaissance and communication satellites. The most powerful Titan variant, called the Titan IIIE, has a Centaur upper stage, and has been used to launch major planetary probes such as Viking to Mars and the Voyager missions to the outer solar system.

THE SATURN V

Of all the rockets used in the space programmes of any nation, without doubt the most impressive is the Saturn V, the launcher designed specifically to send men to the Moon. Once the goal of a

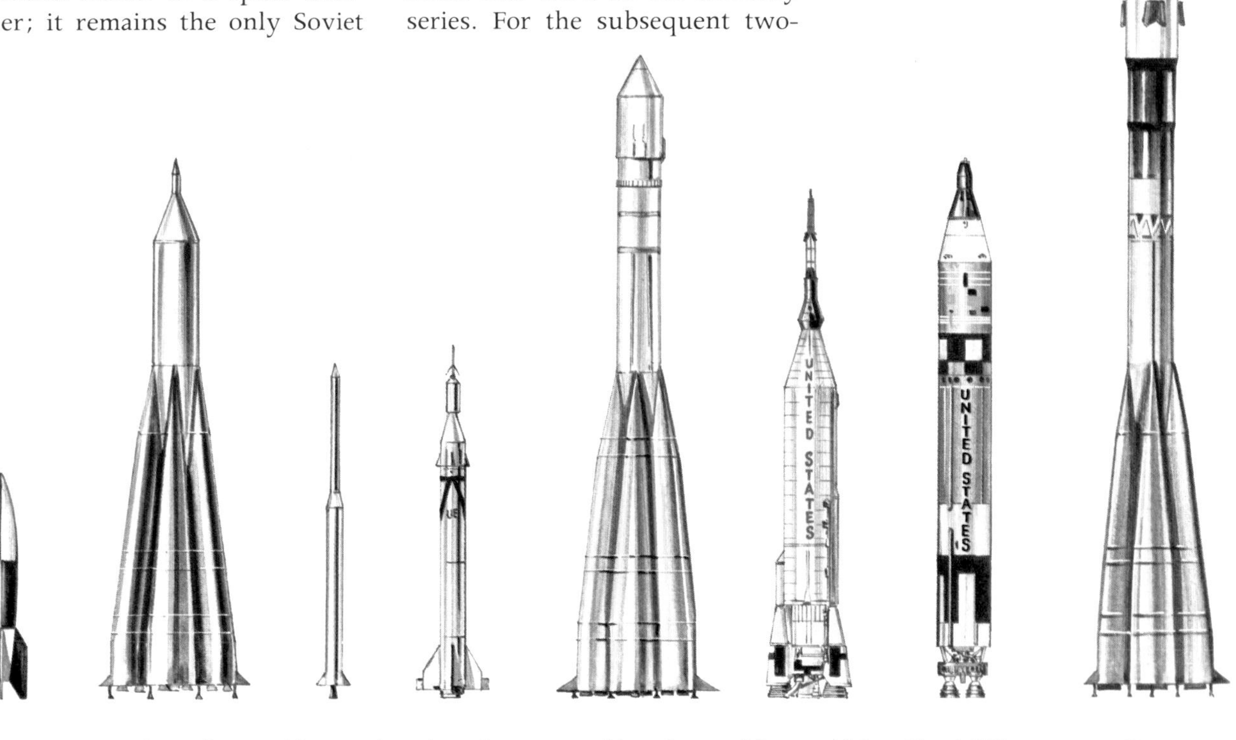

lunar landing had been set in 1961, the United States looked to Wernher von Braun and his group at Huntsville to design and build the rockets to accomplish the task. At the Marshall Space Flight Center, as their headquarters now became called, von Braun's team began work on a series of rockets for manned spaceflight.

First came the Saturn I, a two-stage vehicle which was used to test-launch dummy Apollo capsules. For actual manned flights into Earth orbit an improved

Above: Wernher von Braun, designer of the Saturn V Moon rocket. The Saturn V is the biggest rocket ever to fly, being 111 m (364 ft) high. The scale diagrams *below* show how it towers over the earlier rockets and over the more economy conscious re-usable Space Shuttle.

The Ariane (48 m or 157 ft high) is an unmanned rocket launcher built by the member countries of the European Space Agency to put satellites into orbit.

Saturn 1B　　Saturn V　　Ariane　　Space Shuttle

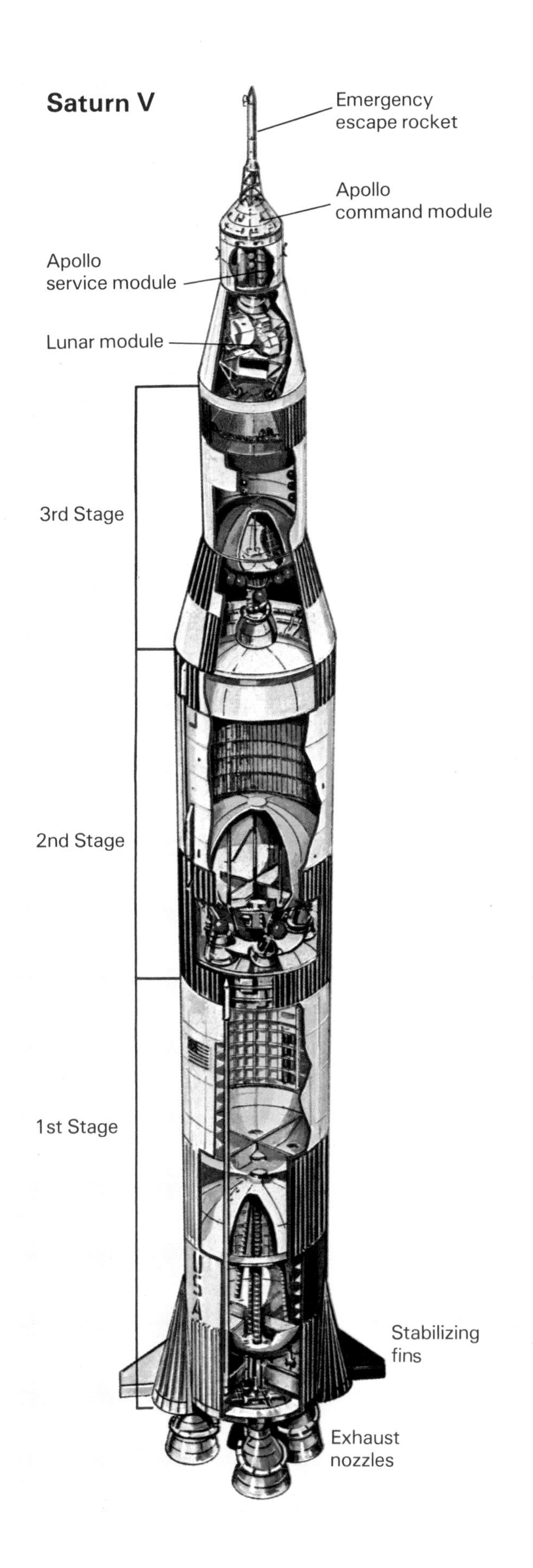

Author Ian Ridpath is dwarfed by a Saturn V at Huntsville, Alabama.

version, the Saturn IB, was developed, capable of launching 18 tonnes into orbit. Saturn IB was used for the Earth-orbital Apollo 7 mission, which was the first manned test flight of the Apollo spacecraft, as well as to send crews to the Skylab space station and to launch the American half of the Apollo-Soyuz international link-up.

But to send Apollo to the Moon required a far larger rocket. For this, the Huntsville engineers developed the Saturn V, the world's largest and most powerful rocket, with five times the power of a Saturn I. It was an unprecedented engineering achievement. Standing 111 m (364 ft) tall, the three-stage Saturn V was able to put over 100 tonnes into Earth orbit, or send over 40 tonnes to the Moon. In addition to the Apollo Moon missions, it was

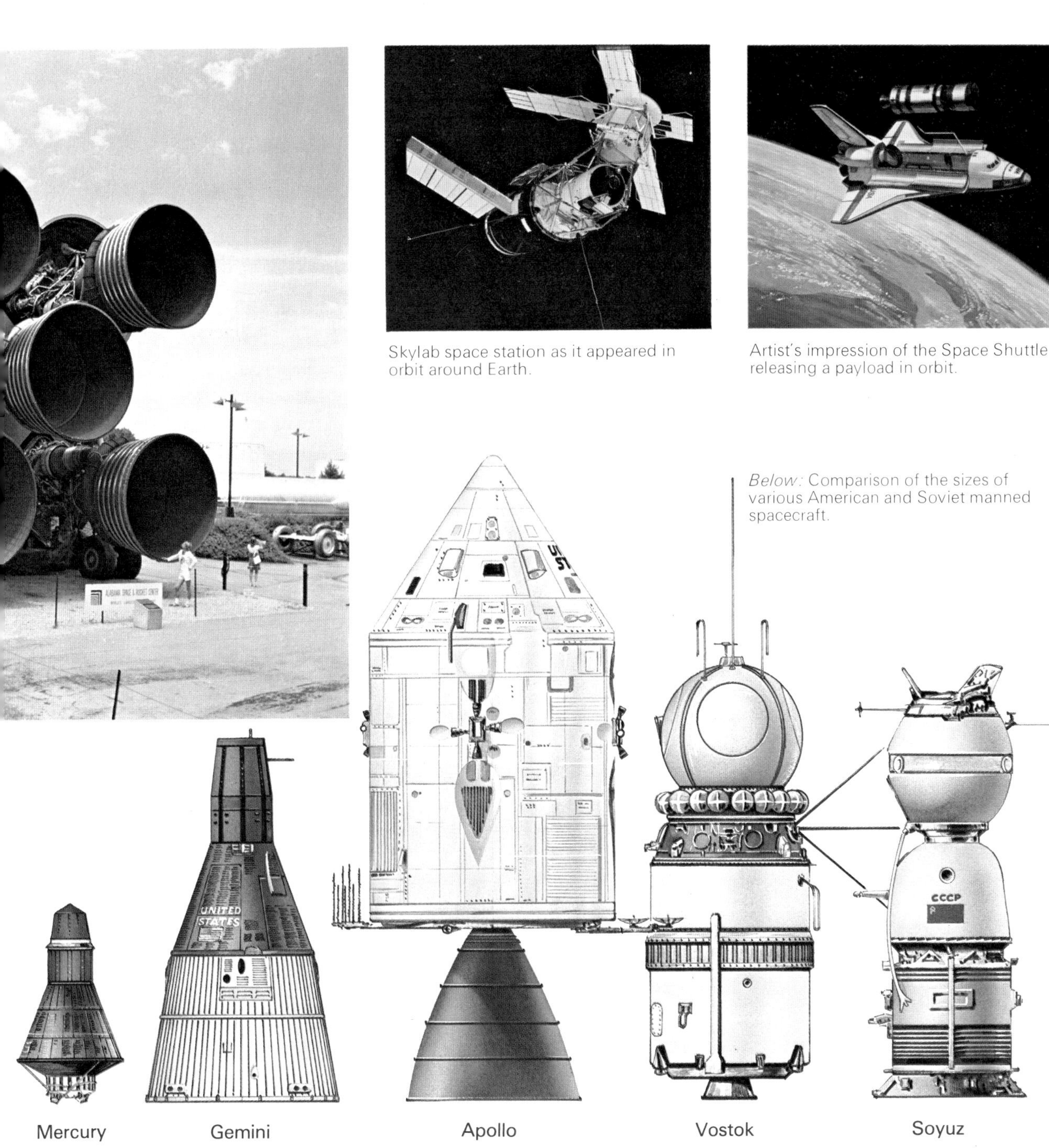

Skylab space station as it appeared in orbit around Earth.

Artist's impression of the Space Shuttle releasing a payload in orbit.

Below: Comparison of the sizes of various American and Soviet manned spacecraft.

also used to launch the Skylab space station, which was a modified version of its own third stage. At the end of the Apollo and Skylab missions, the Saturn series of rockets was pensioned off; we shall never see their like again. Their place has been taken by the winged Space Shuttle which can fly back to Earth and be re-used, thus making it more economical than disposable rockets.

The Soviet Proton launcher lies midway in performance between the Saturn IB and the Saturn V. There have been rumours that the Soviet Union once attempted to fly a secret super-booster bigger even than the Saturn V, presumably for use in a now-abandoned lunar programme. This alleged monster was said to have three stages and be able to send as much as 60 tonnes to the Moon. According to unconfirmed reports, two unsuccessful test launches were made in 1971 and 1972, after which the project was suspended. Perhaps we may see this rocket reappear as part of a Soviet plan for launching giant space stations, or even to send men to the Moon.

3 THE SPACE RACE

On October 4, 1957, a rocket thundered into space from the Soviet launch site near Tyuratam, north east of the Aral Sea. Ten minutes later, the world's first artificial satellite, Sputnik 1, was in orbit. Its launch sparked off a major international rivalry known as the Space Race, as the United States fought to overcome this blow to its prestige while the Soviet Union strove to keep its propaganda lead.

SPUTNIK, THE FIRST MAN-MADE SATELLITE

Sputnik (from a Russian word meaning satellite) was an aluminium sphere 58 cm (23 in) in diameter weighing 83 kg (183 lbs)–as much as a full-grown man. The very weight of the satellite was enough to scare the US, which was incapable of orbiting such large payloads; here was impressive evidence of the Soviet lead in rocket power, and it aroused fears of bombs raining from the skies on American homes.

As Sputnik orbited Earth every 96 minutes, in an elliptical orbit ranging from 215 to 940 km (133 to 584 miles) high, its radio transmitters emitted an eerie 'bleep-bleep' sound. Contrary to popular belief, Sputnik 1 was not packed with scientific equipment and made no major discoveries while in orbit. Rather, it was an engineering test satellite–proof that man-made moons could successfully orbit Earth.

How a Satellite Keeps Up

Perhaps the most common question at the time was: what keeps it up?

The answer is that the effect of Earth's gravity, which tends to pull the satellite downwards, is counterbalanced by the satellite's forward motion, like a weight twirled rapidly at the end of a string. If the satellite is moving fast enough–and this is the importance of the minimum 8 km (5 miles) per second needed to get into orbit, as described in the previous chapter–it will fall in an endless curve around Earth, continually retracing its own path in space while Earth turns underneath it. Once successfully in orbit, a satellite needs no extra rocket power to keep it up.

Four different paths for a spacecraft, depending on its speed when injected into orbit at point P. In orbit (A) the speed is exactly right to counteract the pull of gravity. In orbits B and C, the speed is greater than necessary, so the satellite moves away from Earth before falling back to its starting point. A spacecraft travelling fast enough escapes from Earth altogether — trajectory (D).

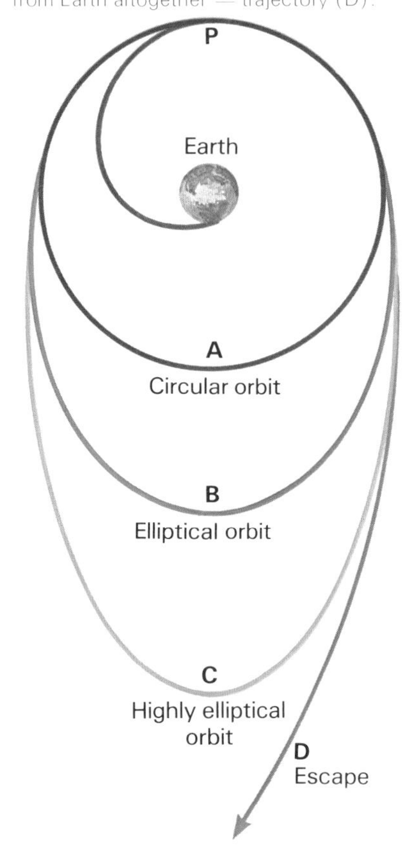

Since Earth's gravity gets weaker with increasing distance, the further away a satellite is, the slower it need move to stay in orbit. For instance, the Moon at a distance of 384 000 km (238 000 miles) completes one orbit of Earth every month, moving at 1 km (0.6 miles) per second. Following these laws, an artificial satellite at the Moon's distance would need to move around Earth every month to stay in orbit; but to get it that far from Earth would require more energy (and hence a bigger rocket) than to place it into a nearer, faster orbit.

If a satellite is moving too slowly to stay in orbit at a given height, gravity will win the uneven struggle and pull the satellite down. If the satellite is moving faster than the required speed for a given altitude, it begins to recede from Earth, slowing down as it does so; then gravity pulls it back to its starting height and its speed increases again to its original value.

The satellite therefore traces

out an orbit that is elliptical in shape. This, incidentally, is why the orbits of most satellites are elliptical rather than perfectly circular; the speed at which a satellite is injected into orbit usually does not exactly match the speed required to counterbalance Earth's gravitational pull at that altitude. For most purposes it does not matter if the orbit is elliptical, so no adjustments are made, but the orbit can be circularized by small, on-board rockets if necessary.

Satellites are usually launched towards the east to take advantage of the rotation of Earth, which gives any departing rocket a free eastward boost. The inclination, or angle, of the orbit to Earth's equator can be adjusted by the direction in which the rocket travels. Some satellites are launched so that they orbit over Earth's poles and are able to pass over the whole of Earth. This type of orbit is used in particular for Earth-monitoring, such as by weather and spy satellites.

A Fiery Death

Sputnik 1 deeply affected all mankind. Millions throughout the world heard its 'bleep bleep' signals on radio and TV, and many satellite spotters saw the top stage of its carrier vehicle orbiting Earth like a brilliant star. Sputnik 1's orbit slowly changed as resistance from the upper atmosphere slowed it down. Tracking the decay of Sputnik 1's path allowed scientists to estimate the density of the upper atmosphere, which turned out to be denser than anticipated, although in laboratory terms it was still close to being a perfect vacuum.

Sputnik 1 fell back to a fiery death in the atmosphere on January 4, 1958. By then the Soviet Union had launched a more impressive package: Sputnik 2, a half-tonne capsule containing the dog Laika, the first living creature to orbit Earth. Sputnik 2 was orbited on November 3, a month after its predecessor, and demonstrated that the power of Soviet rockets was even greater than the West had feared. Laika was kept alive in her pressurized capsule for a week, while reports on her condition were automatically radioed back to doctors on Earth. But Laika was fated to die in space. With no way to bring her safely back, her life was quietly ended. Sputnik 2, with its now-deceased passenger, burned up in the atmosphere on April 14, 1958. Her sacrifice was not in vain; she had shown that living things could travel safely in space.

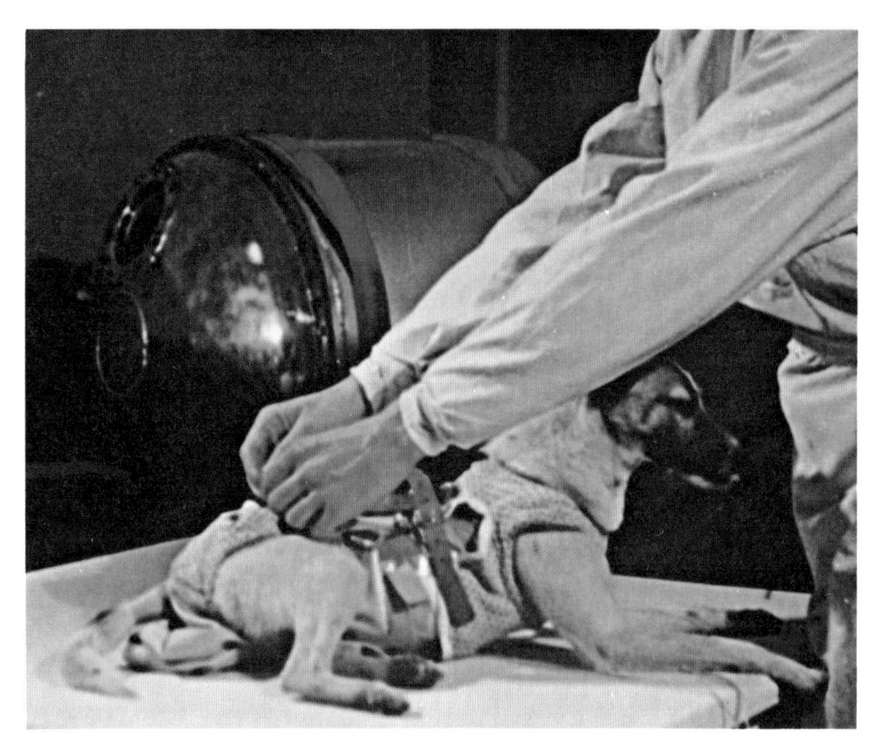

Above: Laika, the first living creature to orbit Earth, is prepared for her trip in Sputnik 2 in 1957.
Below: Viking, an American sounding rocket, was a forerunner of the Vanguard.

THE US SPACE PROGRAMME

Reaction in the United States to these stunning blows by the Soviet Union was swift and strong. The nation that regarded itself as the world's technological leader, and which had publicly announced it would launch the world's first artificial satellite, now found itself upstaged by the Soviet leap into space. Top priority was given to space activities.

There was a good deal of catching up to do. Project Vanguard, America's attempt to place a satellite in orbit, suffered an embarrassing flop a month after Sputnik 2 when the Vanguard rocket settled back on its launch pad and exploded, two seconds after ignition. A second Vanguard launch attempt in February 1958 proved equally unsuccessful.

Explorer

By then, the United States had notched up its first success with Explorer 1, launched from Cape Canaveral by a Juno 1 rocket on January 31, 1958. Explorer 1 weighed a mere 14 kg (30 lbs), but its scientific significance was far greater than its weight. Orbiting every 114 minutes between 356 and 2550 km (220 and 1585 miles) above Earth, Explorer 1 carried scientific equipment that detected the now-familiar Van Allen radiation belts around Earth, named after the American physicist James Van Allen (born in 1914) who devised the experiments on the satellite. Subsequent Explorer satellites confirmed the existence of these belts, which consist of atomic particles from the Sun trapped in Earth's magnetic shell or magnetosphere, which is the volume enveloped by Earth's magnetic field stretching out into space.

Numerous satellites have been launched to study the magnetosphere. The doughnut-shaped Van Allen belts are wrapped round Earth's equator at altitudes around 3000 and 22 000 km (1860 and 12 430 miles); the inner belt contains mostly protons, while electrons dominate the outer belt. The magnetosphere acts as a buffer between Earth's upper atmosphere and the so-called solar wind of atomic particles streaming outwards from the Sun. Occasionally, at times of high solar activity, atomic particles spill over from the Van Allen belts, causing upper atmosphere disturbances such as radio blackouts and the colourful glows known as aurorae.

Explorer 1 had the last word over the early Sputniks; it remained orbiting Earth for more than 12 years (although it stopped transmitting after four months), finally re-entering the atmosphere on March 31, 1970.

Vanguard

Vanguard 1, the second US satellite and the first success in the

Van Allen radiation belts around Earth.

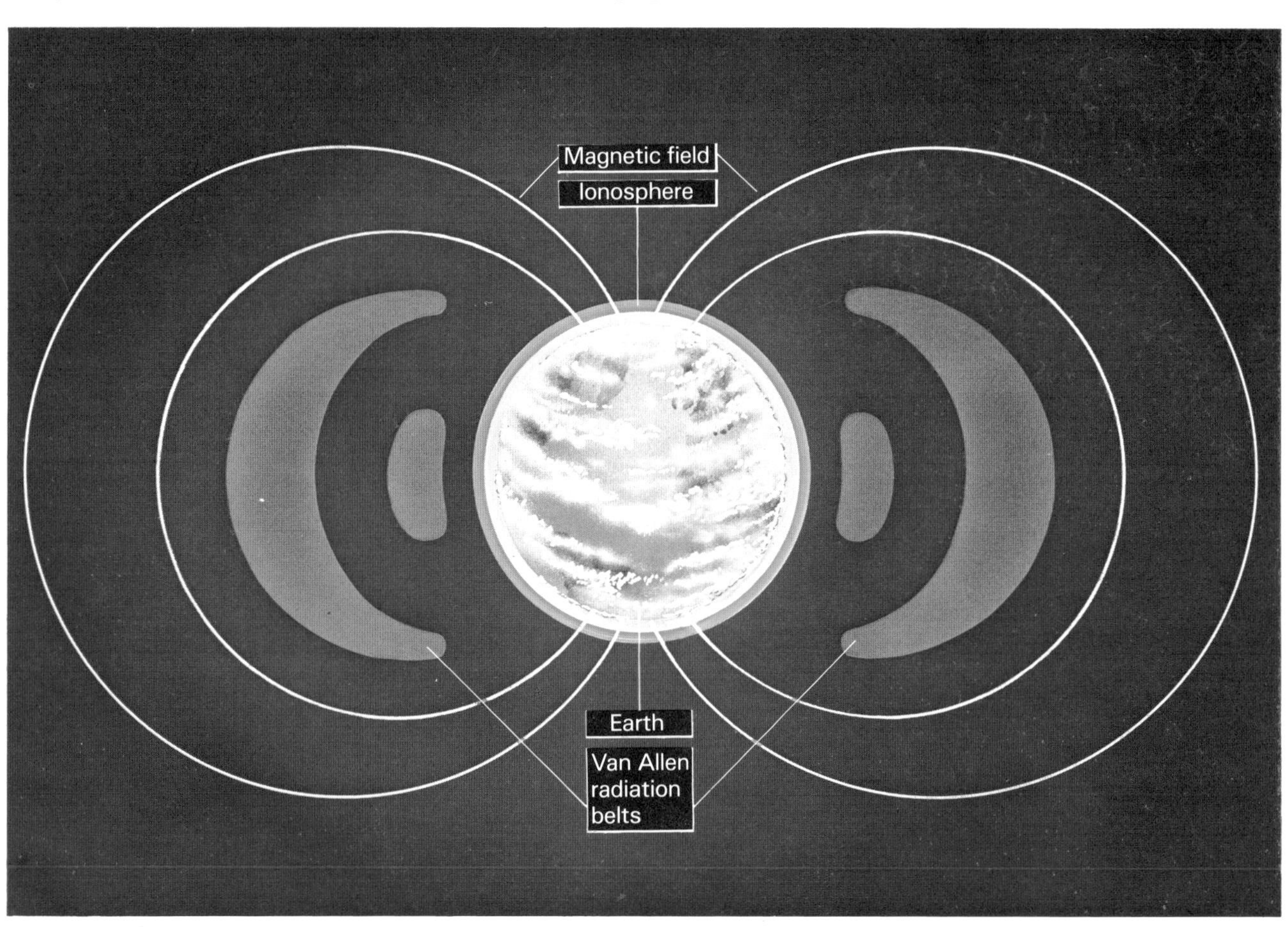

Launch of the American Vanguard 2 satellite in February 1959

Vanguard programme, was launched on March 17, 1958 and proved itself just as scientifically valuable. Tracking of this 1.5-kg (3.3-lb) sphere as it orbited between 650 and 4000 km (400 and 2480 miles) above Earth revealed that our planet is not perfectly spherical but slightly pear shaped. Since Vanguard 1 orbits so high, there is very little air resistance to slow it down, and it is expected to remain in space for thousands of years–unless it is recovered to be put on display in a museum, as seems likely.

Vanguard 1 was a grapefruit-shaped sphere, while Explorer 1 was a long, thin cylinder attached to the small top stage of its launch rocket. Satellites come in all shapes and sizes, though spheres and drums are the usual configurations. The main restriction in a satellite's size and weight is set by the power and diameter of the launch rocket. At launch, the satellite is protected by an aerodynamic shroud; once above the densest part of the atmosphere, this shroud falls away. In the near-airlessness of space, a satellite needs no streamlining–hence the often bizarre configurations of aerials, instrument booms and solar panels sticking out.

Some early spacecraft were powered by batteries but most are powered by solar cells, which turn the energy of sunlight into electricity. Often these cells are studded on the satellite's outer skin, but they can also be mounted on panels. Some satellites and probes carry atomic power plants, which are independent of sunlight; but these can be dangerous if they accidentally crash on Earth, as happened in January 1978 with the Soviet Cosmos 954.

MORE SOVIET SATELLITES

Explorer and Vanguard snatched back some prestige for the United States, but not for long. In May 1958 the Soviet Union launched Sputnik 3, an instrument-packed satellite weighing an astounding 1.3 tonnes. Sputnik 3 confirmed the existence of the Van Allen belts.

Sputnik 3, which was launched by the Soviet Union in May 1958.

The Soviet Union's lead in rocketry made itself felt again as attention turned towards the Moon. Three attempted American Moon shots had failed by January 1959 when the Soviet Union's Luna 1 broke free from Earth's gravity en route for the Moon. It missed by 6000 km (3730 miles) and went into orbit around the Sun like an artificial planet. A second attempt in September was more successful when Luna 2 hit the Moon. For the first time, a man-made object had reached the surface of another body in space. Before Luna 2 crashed, its instrument readings showed that the Moon had no detectable magnetic field.

The Far Side of the Moon

On October 4, 1959, two years to the day after the launch of Sputnik 1, an even more important probe set off into space. This was Luna 3, whose mission was to photograph the far side of the Moon–the side that had never been seen by mankind, because the Moon keeps one face turned permanently towards Earth.

Luna 3's pictures, although poor by modern standards, created a sensation. They showed that the far side of the Moon was different from the familiar near side. There were few dark areas, unlike the visible hemisphere on which vast lowland plains produce the familiar man-in-the-Moon effect. Instead, the Moon's far side consisted almost totally of bright, crater-scarred highland. Later investigations were to show that this is due to the fact that the lunar crust is a few kilometres (or miles) thicker on the far side, so that dark lava from inside the Moon has been unable to spill on to the surface.

By the end of 1959, the United States had launched 18 spacecraft to the Soviet Union's six. Among

Above: Some of the Russian Luna satellite series orbited the Moon. Luna 3 was the first satellite to photograph the far side of the Moon.

these US launches were further members of the Explorer series, which continued to study radiation around Earth and even sent back the first crude pictures of Earth from space. The Air Force began experimenting with a series of recoverable spy satellites called Discoverer. Two intended lunar probes under the name of Pioneer fell short of their goal, but soon the Pioneer series was being used to examine conditions in interplanetary space, particularly to monitor outbursts on the Sun from different vantage points in the solar system. Also around this time, the first prototype satellites for communications, weather and navigation purposes were being orbited by the US (see Chapter 5). The main potential uses of space had already been mapped out.

Although the US was leading in numbers of launches, the propapaganda impact of the Soviet space successes had been far greater. In May 1960 the Soviets launched the 4.5-tonne Sputnik 4, which turned out to be a test flight of the Vostok spacecraft in which Soviet cosmonauts would eventually fly. It was fortunate that Sputnik 4 was unmanned; when the time came to fire its retro-rockets, which would slow the capsule down so that it could return to Earth, the craft was facing the wrong way. Instead of returning to Earth, the craft was boosted into a higher orbit. It remained in space for five months before eventually burning up in the atmosphere.

The next Vostok test flight, disguised under the name Sputnik 5, proved more successful. In August 1960 two dogs, Belka and Strelka, were sent into orbit for one day, returning safely to Earth to become canine celebrities as the first living things to return from orbit. Even so, not all the problems had been ironed out. A repeat of the Sputnik 5 flight, Sputnik 6, carried another two dogs, who perished when the spacecraft entered the atmosphere at the wrong angle and burned up.

Weightlessness

In the United States, plans were being laid for the launching of men into orbit, but inevitably at this early stage of the space race, they lagged behind the plans of

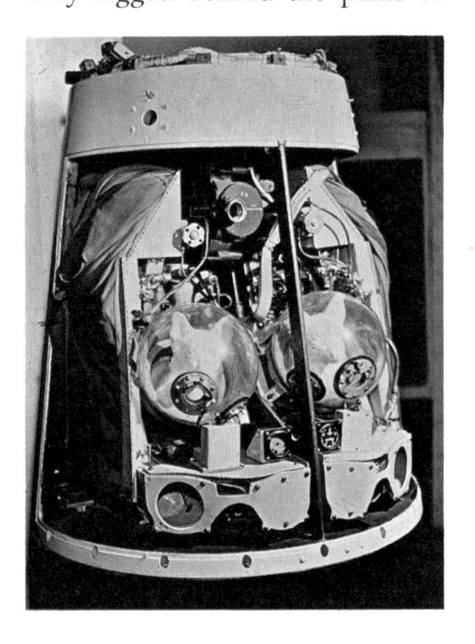

Above: Model of the capsule in which the dogs Belka and Strelka orbited Earth.
Below: The Moon's rugged far side, photographed by Apollo astronauts.

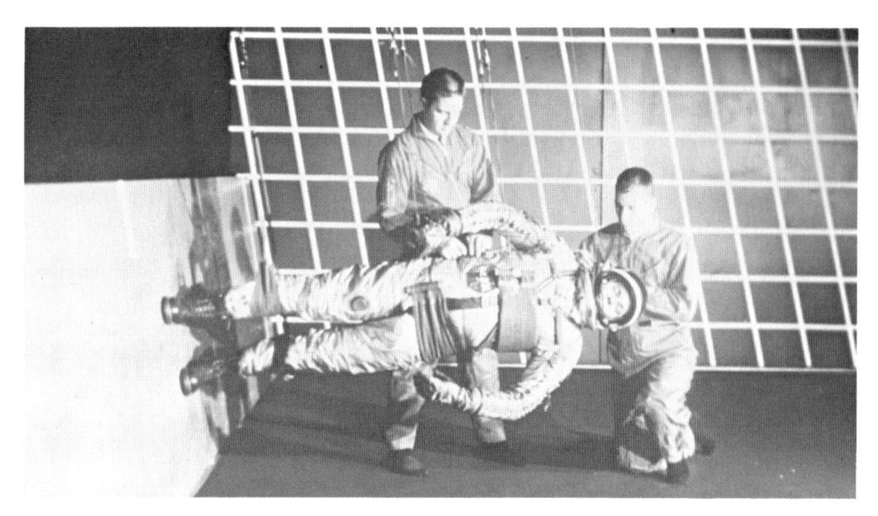

Top left: The natural buoyancy of water helps simulate weightlessness. Here, astronaut Edwin Aldrin trains in a water tank for a walk in space.
Above: An astronaut is suspended horizontally in a special harness to simulate the feeling of walking in the low-gravity conditions on the Moon's surface.

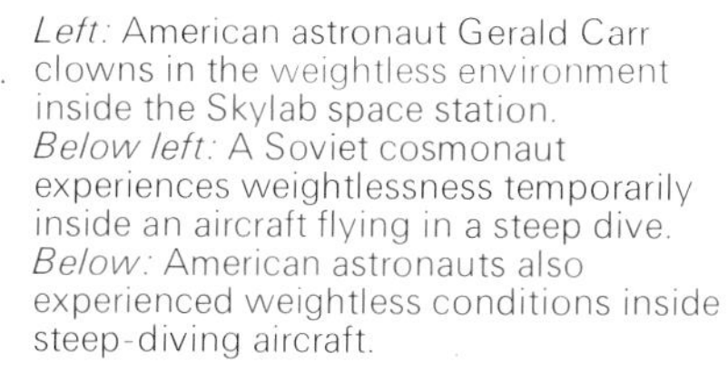

Left: American astronaut Gerald Carr clowns in the weightless environment inside the Skylab space station.
Below left: A Soviet cosmonaut experiences weightlessness temporarily inside an aircraft flying in a steep dive.
Below: American astronauts also experienced weightless conditions inside steep-diving aircraft.

the Russians. Project Mercury, as this first American manned programme was called, was intended to place a man into space, test his ability to function there, and return him safely to Earth.

At that time no one knew if a man could survive the conditions in orbit where his body would become weightless, or if he could withstand the strains of take-off and re-entry when his body would be subjected to forces that would make him feel many times his normal weight.

A word needs to be said about weightlessness. A spacecraft orbiting Earth is not outside Earth's gravity, yet it (and any occupants) are weightless. How can this be? The answer lies in the fact that the spacecraft's forward motion counteracts the pull of gravity, as discussed at the start of this chapter. Earth's gravity pulls the spacecraft towards it, and the spacecraft responds. But at the same time the spacecraft is moving forwards.

If Earth were flat, the spacecraft would inevitably hit its surface. But Earth is spherical, so the ground seems to fall away from under the spacecraft as it moves. The combination of the forward motion of the spacecraft and the downward pull due to gravity means that the spacecraft moves in a curve around Earth. A satellite is, for practical purposes, falling endlessly around Earth, which is why conditions in orbit are also known as 'free fall'. And anything that is falling has no weight.

You can demonstrate the truth of this (though I'm not suggesting you actually do it!) by jumping off a high building while standing on bathroom scales. During the descent the scale would read zero because your body would not be exerting any force on it. Only once you touched the ground would the sensation of weight return.

Among the fears expressed for weightless astronauts were that they would become disoriented or suffer heart irregularities. In the event, the worst fears proved groundless, but prospective astronauts were deliberately selected for peak fitness. They were put through a strenuous training programme which included being whirled in centrifuges to make them feel many times their normal weight, thereby simulating the forces of take-off and re-entry. And there were test flights by animals–dogs in the case of the Russians, chimps for the Americans.

Interior of a Soviet Vostok spacecraft showing a porthole and some of the controls and instruments that faced the cosmonaut.

Mercury and Vostok

The one-man Mercury spacecraft was conical in shape, 1.9 m (6.2 ft) across at its widest and 2.9 m (9.5 ft) long. It weighed 1.3 tonnes against the 2.4 tonnes of the Vostok sphere. Part of the reason for the weight difference is that American spacecraft use a pure oxygen atmosphere at low pressure, as against a normal Earth atmosphere for Soviet spacecraft; hence the American craft can be made lighter, although Mercury was in any case smaller than Vostok.

Inside Mercury, the astronaut sat clad in his spacesuit on a couch contoured to his body shape, his head touching one wall and his feet braced against the other. It was a tight squeeze, and not for the claustrophobic. In front of him was an instrument panel, and in his hand he held a joystick with which he could control the orientation, or attitude, of the spacecraft while in orbit. At the spacecraft's blunt end was the heat shield, designed to resist the fierce temperatures as the capsule blazed into the atmosphere during re-entry. (Manned spacecraft reenter blunt end first to slow down before their parachutes open). Strapped to the back of the Mercury heat shield were the retro-rockets which fired to put the spacecraft on course back to Earth.

By contrast, the design of the Soviet Vostok, which like the launch rocket was kept secret for years, was a sphere 2.3 m (7.5 ft) in diameter, coated with heat-resistant material. A small instrument unit behind the sphere,

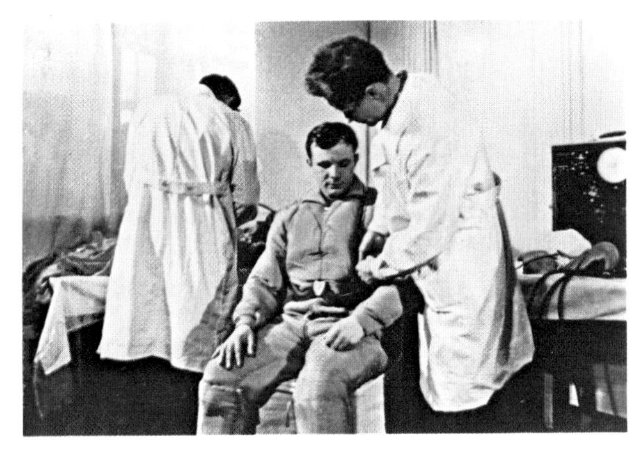

jettisoned (cast off) before re-entry, contained gas bottles to provide air for the cosmonaut, as well as electrical batteries and retro rockets.

The cosmonaut inside had very little control over the movement of his spacecraft. Despite its greater size and weight than Mercury, Vostok was little more than an unmanned satellite with a passenger. The cosmonaut sat in an ejector seat which he could use to parachute to safety in the event of a launch emergency; Mercury spacecraft had a launch escape tower powered by small rockets which would have pulled the entire spacecraft free had the launch rocket malfunctioned. When landing, the Vostok cosmonauts were ejected from their capsules and parachuted down separately. Mercury capsules splashed down in the sea with their pilots safely aboard.

Above left: Soviet space doctors prepare Yuri Gagarin for his pioneer spaceflight in April 1961.
Above: A Vostok spacecraft being readied for launch. The Vostok itself was a sphere, but at launch was protected by a white cover, as shown here.

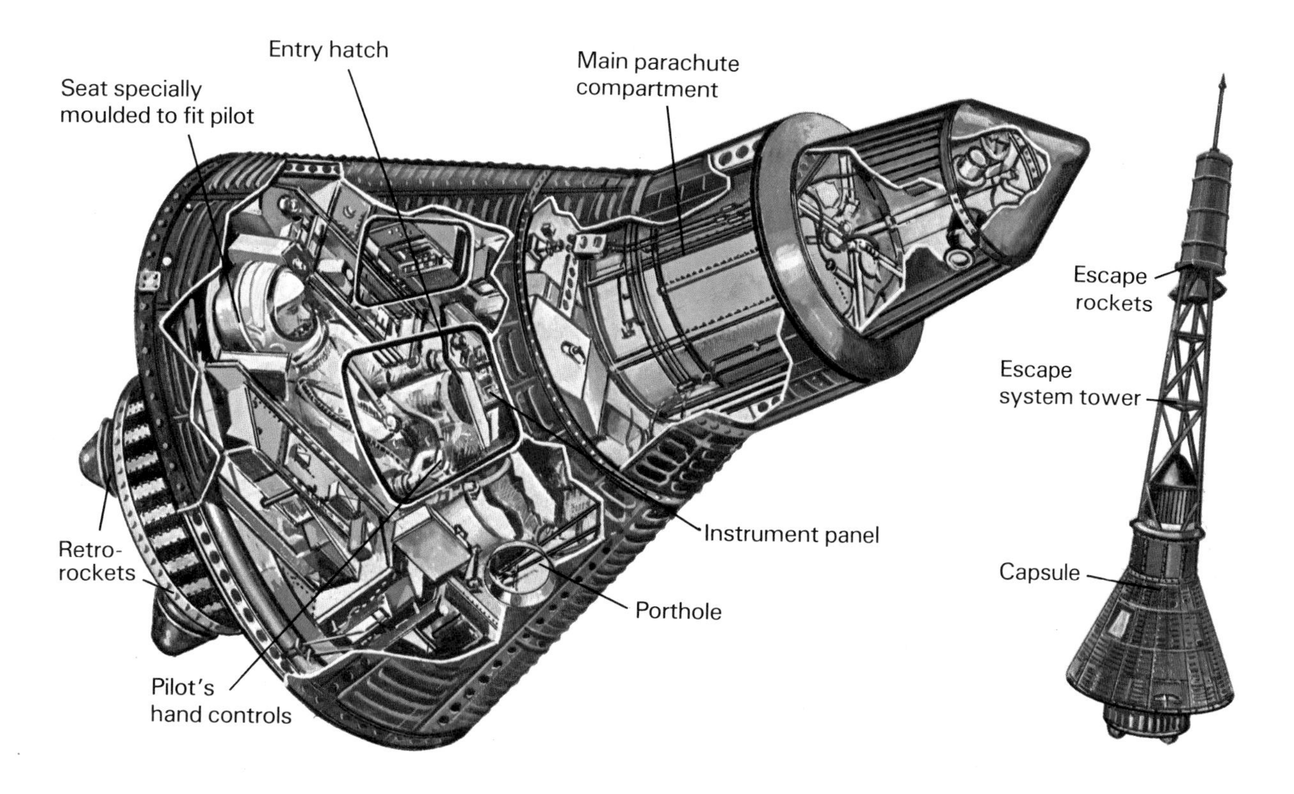

A cutaway diagram of the one-man Mercury spacecraft. It was a tight squeeze for the astronaut, and not for the claustrophobic.

Far left: Yuri Gagarin, the world's first spaceman, seated in his Vostok spacecraft. *Left:* Heavily retouched Soviet picture of Gagarin's launch. For many years the Soviet Union concealed the true design of their space launch rockets. *Above:* Gagarin alights from a recovery aircraft after his pioneer spaceflight.

THE FIRST SPACEMAN

In spite of their design differences, both manned spacecraft were successful. And it was the Russian Vostok that was first into action. After two more test flights in the Sputnik series in which dogs were orbited and successfully recovered, Soviet space planners decreed that it was time for the first human to orbit the Earth. His name was Yuri Alekseyevich Gagarin (1934–1968).

On April 12, 1961, Gagarin was rocketed into orbit from the Soviet space base at Tyuratam. After completing one orbit, the automatic controls of his spacecraft brought him back to Earth as scheduled. To everyone's relief he returned safe and sound. He practised eating, drinking and writing while in orbit, and reported no difficulty in adapting to weightlessness.

He said of his flight: 'When I became weightless I felt perfectly well. Objects floated in the cabin. I myself was not sitting in my chair as before–I hung in the air. For the first time I saw the spherical shape of the Earth. In space the Sun is many times brighter than on Earth.'

His entire flight, from lift-off to touchdown, lasted 108 minutes. But its significance was far greater than that. Gagarin's single orbit of Earth caused a sensation as great as the first Sputnik's launch.

Again the United States was stunned. It was conducting its space programme in the full glare of international publicity while the Russians paradoxically profited from their policy of secrecy and surprise. Gagarin became a world hero, and the US manned space programme had to operate in the shadow of his triumph. Yet fate decreed that Gagarin would never make another spaceflight. In 1968, while training for a mission in the multi-man Soyuz series, he was killed in an aircraft crash.

America's first two manned shots were less ambitious than the orbital flight of Gagarin. They were so-called sub-orbital flights in which the astronauts were sent on simple up-and-down trajectories beyond the atmosphere and back again. To do this, the Mercury capsule was placed atop a modified Redstone missile.

First to fly was Alan B. Shepard (born in 1923). On May 5, 1961, two weeks after Gagarin, he was lobbed 490 km (304.5 miles) into the Atlantic from Cape Canaveral, reaching a maximum altitude of 187 km (116 miles). His entire flight took just over 15 minutes, during which he was able to make observations of Earth and the sky and control the attitude of his spacecraft with tiny rocket jets. Shepard confirmed what Gagarin had shown: that there was no obstacle to man functioning successfully in space.

A second sub-orbital flight, this time by Virgil I. Grissom (1926–1967), nearly ended in disaster when the hatch of his spacecraft accidentally blew off after splashdown. The spacecraft sank, but fortunately Grissom was rescued by helicopter.

Left: Preparations at Cape Canaveral for Virgil Grissom's sub-orbital flight in July 1961.
Above: Grissom squeezes into his Mercury capsule.
Below left: Alan Shepard, America's first spaceman, being winched aboard a helicopter after splashdown and, *inset*, Shepard talking to doctors.

Although these were impressive achievements, they paled into insignificance beside the pioneer flight of Gagarin.

For the next stage in its Mercury programme, the United States prepared to launch men into orbit, atop the more powerful Atlas booster. But, before this could happen, the Soviet Union produced another surprise.

Space Sickness

On August 6, 1961, cosmonaut Gherman S. Titov (born in 1935), who had been Gagarin's backup, made a 17-orbit flight lasting just over a day–an achievement in its own way as daring as Gagarin's. Titov ate, worked, and slept in orbit. His flight was not trouble-free, for he suffered a malady that was to affect a number of subsequent spacemen–space sickness. Slight nausea troubled Titov a few hours into his flight, particularly when he moved his head rapidly.

So-called space sickness turns out to be one of the main problems

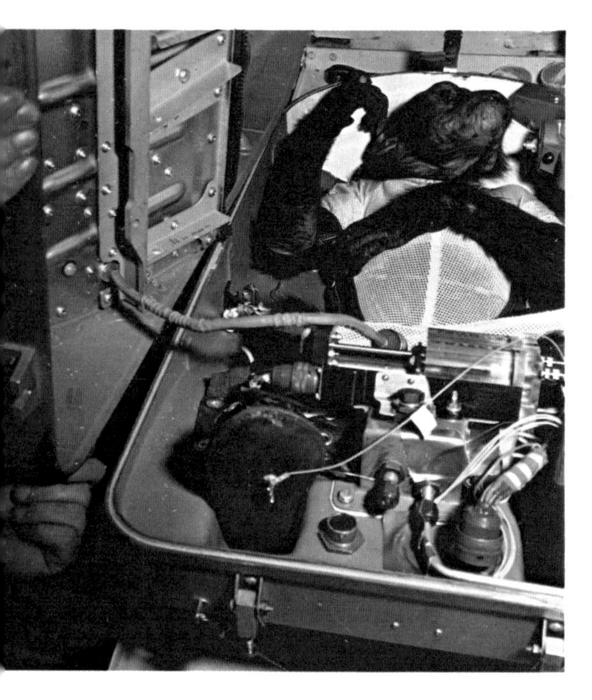

Above: Chimpanzee Enos made a three-orbit flight of Earth in preparation for the orbital mission of John Glenn in 1961. Enos is here being fitted into his pressure couch.
Above right: Glenn entering his Mercury spacecraft.

of adjustment to weightlessness. It seems to be caused by sensors in the inner ear which normally tell the body which way up it is; in weightlessness, these sensors are unable to function correctly. Like motion sickness on Earth, it affects different people to different degrees. The problem proved particularly bad in flights of larger spacecraft, notably the Skylab space station where astronauts had more room to move around and were thus more likely to become disoriented. Fortunately, the problem passes away after a few days at most.

Titov's flight showed that man could survive more than fleeting bouts of weightlessness, although the true limits of human endurance in space remained unknown. Only more extended missions, to be made later, would reveal if there were any medical barriers to long-term spaceflight.

Americans in Orbit

By February 1962, following a test flight by a chimpanzee called Enos, the time had come for the United States to place its first man in orbit. He was John H. Glenn (born in 1921). As the world watched in suspense, Glenn was launched into orbit atop an Atlas booster, making three revolutions of Earth before returning. His flight proved eventful for unforeseen reasons.

Firstly, he reported mysterious 'fireflies' outside his capsule, which seemed to be following him. Later, it was realized that these glowing particles were drops of moisture from vents on the spacecraft.

Glenn found he could distinguish details on Earth with great ease, such as ocean currents, cities and land formations. He used his hand controls to guide the Mercury capsule during flight, showing that a man could safely pilot a spacecraft in orbit.

What looked to be a frightening problem for Glenn developed on the third and final orbit, when ground controllers received a signal indicating that the Mercury capsule's heat shield might be loose. Fortunately, the signal turned out to be false, and Glenn returned safely. Had the heat shield really been loose, it could have torn away during re-entry, thus making Glenn the world's first space fatality. Glenn was a national hero after his flight but never went into space again, opting to become a US senator instead.

Three months later, Glenn's backup, M. Scott Carpenter (born in 1925), made a repeat performance of the flight, undertaking some simple experiments and observations in orbit. Technical troubles led to Carpenter overshooting the landing site by 400 km (248 miles), but he was safely recovered. It looked as though the United States was catching up in the space race. But, while they prepared for longer flights, the Soviet Union sprang a new space spectacular, this time involving *two* Vostok spacecraft.

First off, on August 11, 1962, was Andrian G. Nikolayev (born in 1929) in Vostok 3. The following day he was joined in orbit by Vostok 4 carrying Pavel R.

Popovich (born in 1930). Together the space twins circled Earth, landing together after four days and 64 orbits for Nikolayev and three days, 48 orbits for Popovich. At their closest the two came within 6.5 km (4 miles) of each other, but there was no chance of a collision. Since Vostok had no manoeuvring engines there was certainly no opportunity for a true space rendezvous. Soviet space doctors were particularly concerned to study the physical effects of spaceflight on the two cosmonauts; they found no lasting effects, although it became clear that a temporary weakening of the cosmonauts' hearts and muscles was taking place in weightlessness.

There were two more flights scheduled in the American Mercury series. In the first, in October 1962, Walter M. Schirra (born in 1923) orbited Earth six times, spending just over nine hours in space. Most ambitious of all, pushing the capabilities of Mercury to the limit, was L. Gordon Cooper's (born in 1927) 22-orbit flight in May 1963. It lasted 34 hours, longer than all the previous Mercury missions combined.

One special feature of Cooper's flight was the observations he made of the ground. Photographs taken by previous astronauts had demonstrated the value of observing Earth and the atmosphere from space. In addition to continuing this photographic programme, Cooper reported seeing astoundingly fine details on the ground by eye, such as roads and railway lines. Later missions confirmed these sightings. Cooper released a flashing beacon from the capsule to judge how easy it was to track objects visually in orbit; this knowledge was necessary for future attempts at docking in space.

American astronaut L. Gordon Cooper trains for his 22-orbit Mercury flight in 1962.

During his flight Cooper tried out various sorts of space food. Some food came as bite-sized chunks, while other food was a paste in a plastic tube which could be squeezed out into his mouth. Cooper's flight was not without its problems. Carbon dioxide built up in his space suit, and he had to control his re-entry manually due to a failure of the autopilot. But splashdown was on target, and

Examples of dehydrated space foods, from left to right: Tuna salad, salmon salad, orange drink, strawberry cubes.

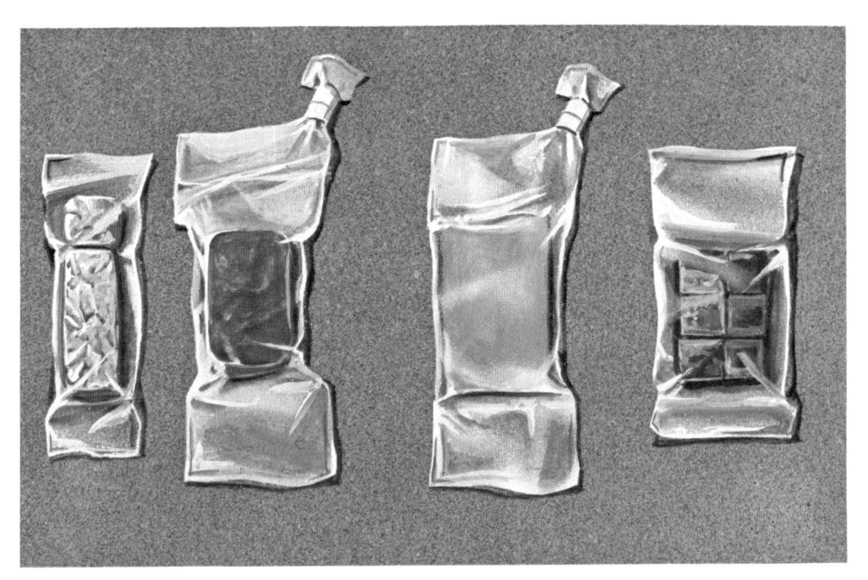

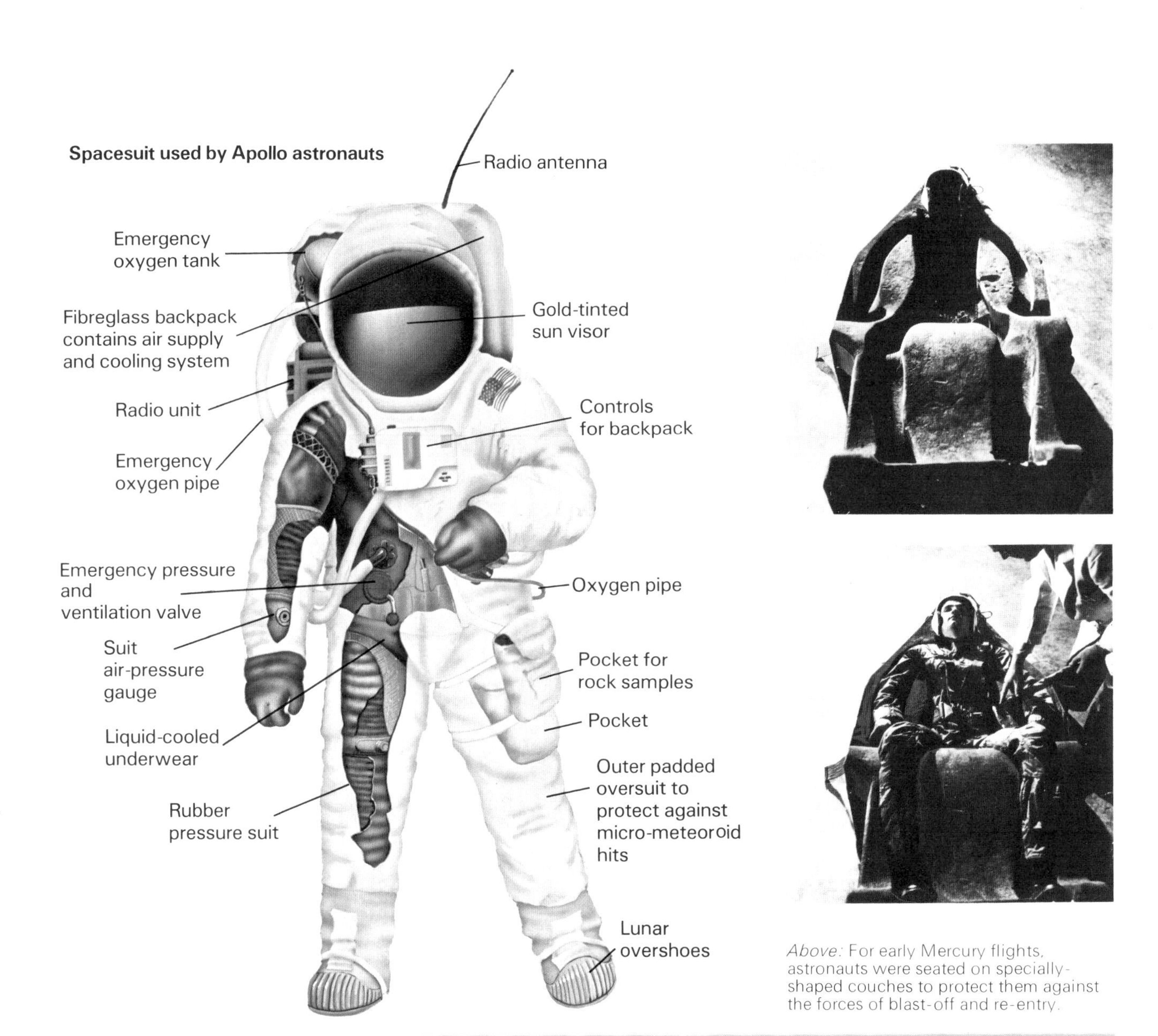

Above: For early Mercury flights, astronauts were seated on specially-shaped couches to protect them against the forces of blast-off and re-entry.

the Mercury programme ended on a note of triumph.

The First Spacewoman

The Soviet Union had another surprise up its sleeve before the end of its Vostok programme. A month after Cooper's flight came another double Vostok launch–this time with a woman, Valentina V. Tereshkova (born in 1937), on board. She made 48 orbits of Earth before returning safely, thus showing that women were as capable of travelling in space as men. However, she never made another flight, and it soon became clear that the Soviet Union had no plans to put more women into space. In November

SOVIET MANNED VOSTOK FLIGHTS

Mission	*Launch Date*	*Remarks*
Vostok 1	April 12, 1961	Yuri Gagarin made one orbit of Earth. First manned spaceflight.
Vostok 2	August 6, 1961	Herman Titov made one-day, 17–orbit flight.
Vostok 3	August 11, 1962	Andrian Nikolayev orbited Earth 64 times, landing on August 15. Joint flight with Vostok 4.
Vostok 4	August 12, 1962	Pavel Popovich orbited Earth 48 times simultaneously with Vostok 3. Landed August 15.
Vostok 5	June 14, 1963	Valery Bykovsky orbited Earth 81 times, landing on June 19. Joint flight with Vostok 6.
Vostok 6	June 16, 1963	Valentina Tereshkova orbited Earth 48 times simultaneously with Vostok 5, landing on June 19. First space woman.

1963 Valentina married fellow cosmonaut Andrian Nikolayev, hero of Vostok 3. The couple became proud parents of a healthy daughter, thus dispelling fears that radiation in space might cause genetic damage. Many other spacemen have subsequently fathered perfectly normal children.

Valentina's comrade in the final Vostok double mission was Valery F. Bykovsky (born in 1934), who piloted Vostok 5 for 81 orbits of Earth, lasting nearly five days. This remains the longest solo spaceflight in history.

As the Mercury and Vostok projects drew to a close, sights were already being set on more ambitious goals. Goaded by the space successes of the Soviet Union, US President John F. Kennedy sought a challenge in which the United States could hope to come first. Among the ideas considered was establishing a manned space station, but this was rejected because it seemed insufficiently dramatic and also because the Russians, with their advanced rocket power, appeared certain to reach the target first. There was only one suitable choice: to place the first man on the Moon.

On May 25, 1961, a mere three weeks after the sub-orbital flight of Alan Shepard, President Kennedy laid down his historic challenge: 'I believe that this nation should commit itself to achieving the goal, before this decade is out, of landing a man on the Moon and returning him safely to the Earth.'

The space race had become the Moon race.

Top right: Valentina Tereshkova, the first woman to fly in space, sits in the cabin of her Vostok spacecraft.
Right: The Moon — the goal that President John F. Kennedy *(inset)* committed the United States to reach before 1970 — and before the Russians.

AMERICAN MANNED MERCURY FLIGHTS

Mission	*Launch Date*	*Remarks*
Mercury-Redstone 3 *(Freedom 7)*	May 5, 1961	Alan Shepard made suborbital flight. First American in space.
Mercury-Redstone 4 *(Liberty Bell 7)*	July 21, 1961	Virgil Grissom made suborbital flight.
Mercury-Atlas 6 *(Friendship 7)*	February 20, 1962	John Glenn orbited Earth three times.
Mercury-Atlas 7 *(Aurora 7)*	May 24, 1962	Scott Carpenter orbited Earth three times.
Mercury-Atlas 8 *(Sigma 7)*	October 3, 1962	Walter Schirra orbited Earth six times.
Mercury-Atlas 9 *(Faith 7)*	May 15, 1963	Gordon Cooper made day-long flight, orbiting Earth 22 times.

4 THE MOON RACE

To land a man on the Moon, the target that the United States set itself in 1961, required more experience in manned spaceflight than provided by project Mercury.

A three-man craft known as Apollo was designed for Moon missions, but first came an intermediate step, project Gemini. In the Gemini series of flights, two men would practice the arts of rendezvous and docking with other spacecraft, and of walking in space—extravehicular activity, or EVA as it was technically known. These techniques had to be mastered in the mid-1960s if the United States was to fulfil President Kennedy's challenge of putting a man on the Moon by 1970. As it happened, project Gemini succeeded spectacularly well and catapulted the United States into the space-race lead.

GEMINI

At first sight, Gemini looked like an enlarged version of Mercury; in fact it was far more sophisticated. The two-man crew sat in ejector seats in a conical capsule called the re-entry module, 2.3 m (7.5 ft) wide at its base. In the event of a launch emergency the crew could eject to safety–there was no launch escape tower atop the Gemini spacecraft. Gemini had an extended nose section which contained radar for making rendezvous with other spacecraft in orbit, and parachutes for return to Earth.

Lift-off of a Titan rocket carrying a two-man Gemini spacecraft.

Behind the dark-coloured re-entry module were two other compartments, painted white. At the back of these was the equipment section which contained fuel cells for generating electricity, and engines for manoeuvring while in orbit. Before re-entry this section was jettisoned to reveal retro-rockets in the second compartment; after firing, this section was also jettisoned to expose the Gemini heat shield which took the full force of re-entry. Altogether, Gemini and its equipment sections weighed 3.6 tonnes, and totalled 5.6 m (18.3 ft) in length. Titan 2 rockets were used to put Gemini into orbit.

The major advance of Gemini was its ability to change orbit for rendezvous and docking with other spacecraft. A miniature on-board computer, in conjunction with the Gemini nose radar, calculated the necessary orbital changes; the computer also aided accurate re-entry and splashdown. Gemini carried sufficient supplies for extended missions which showed that men could stay safely in space long enough to reach the Moon and return.

WALKING IN SPACE!

But before the much-heralded Gemini series could get under way, the Soviet Union sprang

another surprise by launching a spacecraft called Voskhod. Voskhod 1, in October 1964, carried a three-man crew for a day; it appeared as though the Soviet Union had jumped a stage in the space race to reach the equivalent of Apollo.

The following March, Voskhod 2 created an even greater sensation. It carried two men, one of whom, Alexei Leonov, crawled out through an airlock to make the first walk in space. He was outside Voskhod for 10 minutes, tethered by a cable, breathing oxygen from a pack on the back of his spacesuit. At the end of his lifeline, Leonov performed acrobatics in orbit. For the first time, a man demonstrated that he could perform in orbit independently of his spacecraft, like a temporary, living Earth satellite.

Leonov's success was reassuring, for some pessimists had warned that a man on his own in open space might become disoriented–warnings similar to

Left: Alexei Leonov, the first man to walk in space, seen during training.
Below: Leonov walking in space.

Below: A cutaway diagram of the Gemini spacecraft, showing part of the interior.

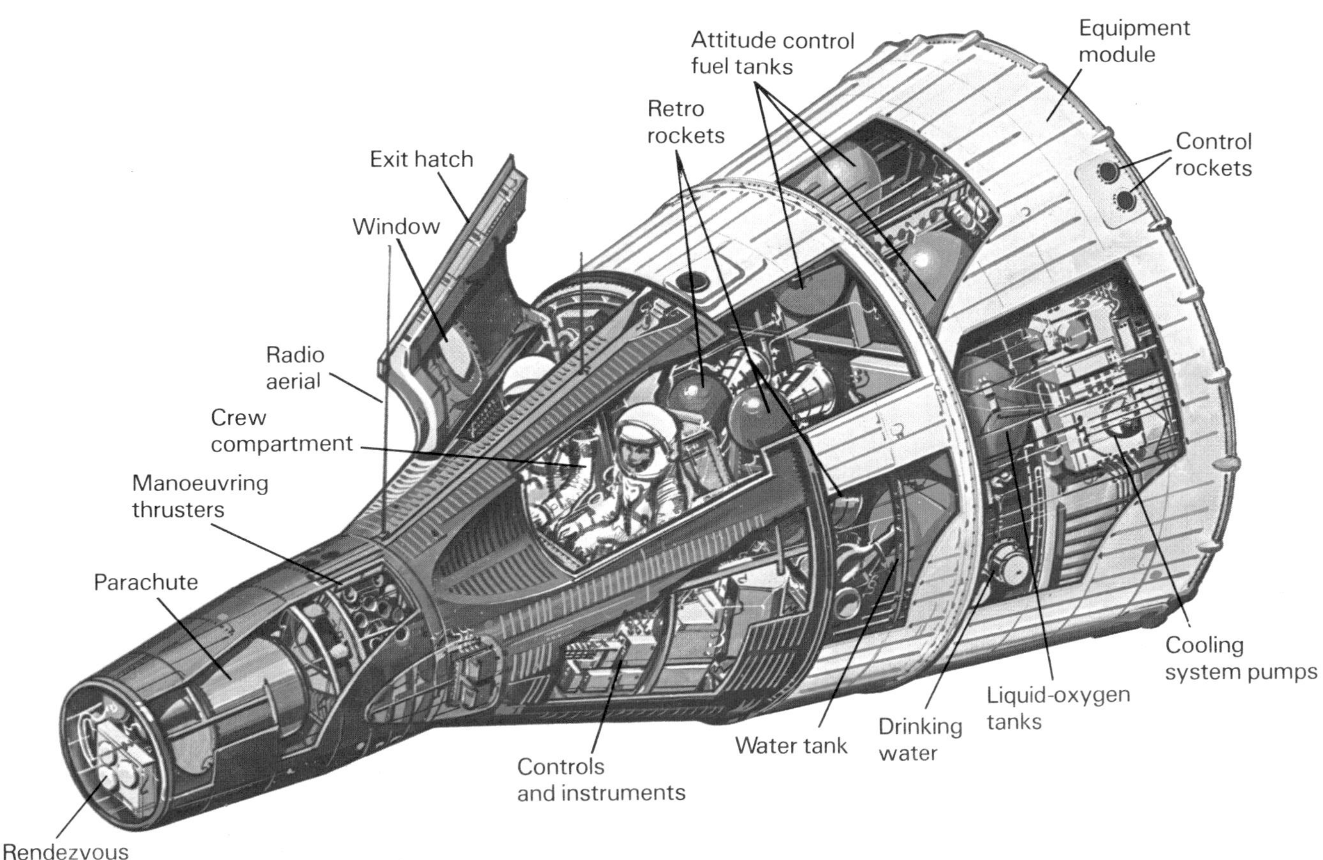

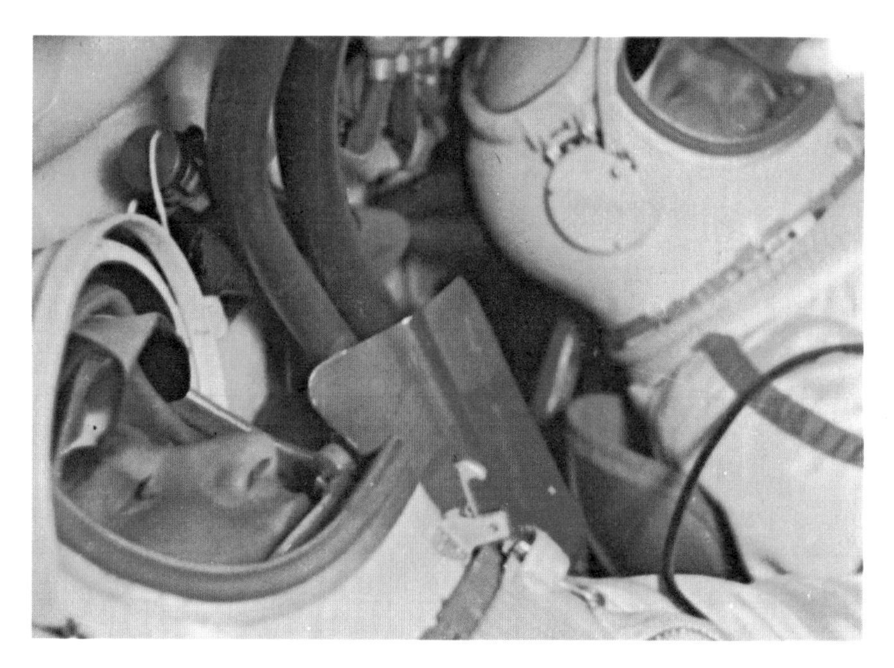

Above left: Soviet cosmonauts Alexei Leonov (background) and Pavel Belyaev in the cabin of their Voskhod spacecraft. Leonov crawled out from Voskhod through the airlock to become the first human to walk in space.
Above: American astronauts James McDivitt and Ed White (background) rehearse for their Gemini 4 flight, during which White became the first American to walk in space.
Right: Ed White photographed by McDivitt during his space walk. He was tethered to the Gemini capsule by a cable, and carried a small jet gun in his hand to help him move in space.
Below: Launch of the Titan rocket carrying Gemini 4 into orbit in June 1965.

the pessimistic predictions before the first man flew in space. But not all Voskhod 2's problems were over when Leonov was safely strapped back in his capsule. The Voskhod's automatic re-entry system malfunctioned and the crew had to land manually, overshooting their intended target by 1600 km (994 miles). They ended up in a snowdrift from which they were rescued several chilly hours later.

Voskhod had an improved landing system, involving the firing of small gas jets just before touchdown, so that the crew could land safely inside the spacecraft. This was just as well because there was no way for them to eject.

Little was said about the design of Voskhod at first. But eventually it became clear that it was only a modified Vostok–a stopgap to upstage Gemini. By removing the bulky Vostok ejector seat, Soviet space designers were able to cram three men into the same-sized sphere as had held one man. Because of the lack of room, the three-man Voskhod 1 crew had to fly without spacesuits. And, without ejector seats, there was no way for them to escape. Had there been a launch failure on either of the Voskhods, the crew could well have been killed.

The airlock through which Leonov crawled in his Voskhod 2 flight turned out to be nothing more than an inflatable concertina-like affair fixed over a hatch, discarded once he was back on board. Voskhod was a dangerous attempt to snatch space propaganda, and it is not surprising that the series was abruptly halted after two flights.

Project Gemini eventually got under way with a three-orbit test flight, the first in which a manned spacecraft changed orbit, which it did by firing its on-board engines. Commander of the flight was Virgil Grissom, veteran of the second Mercury sub-orbital shot, who therefore became the first man to undergo the rigours of two space launches. It was a relatively low-key beginning to a series that was later to achieve significant space records and space 'firsts'.

Gemini Achievements

The real fun began with the second manned mission, Gemini 4 (the first two had been unmanned

tests). It was the longest US space mission to date, lasting a full four days. During the flight, astronaut Edward H. White (1930–1967) made a walk in space. Gemini did not have an airlock. Instead, the spacecraft hatch was opened and one astronaut clambered out. When he returned, the hatch was fastened and the capsule repressurized.

Astronaut White remained outside Gemini 4 for 21 minutes, twice as long as Leonov's space walk. He was supplied with oxygen through an umbilical cord which also tethered him to the spacecraft. He was able to manoeuvre with a hand-held gun which fired small jets of gas. He so enjoyed the experience that he was reluctant to get back in the spacecraft!

Following this success came a new space endurance record–eight days in orbit by L. Gordon Cooper, his second flight, and astronaut Charles Conrad (born in 1930), his first flight. (Later Conrad was to become one of America's most experienced spacemen.)

Their Gemini 5 orbited Earth a record 120 times, making them the most travelled men in history–although this feat was to appear trivial compared with later space efforts.

Cooper proved that the fine details he had seen on Earth's surface during his Mercury mission had not been misperceptions. Both astronauts in Gemini 5 agreed that even wakes of ships at sea and the contrails of aircraft could be spotted by eye from orbit. The Gemini 5 crew made an extensive series of photographs of Earth and these findings served as a basis for later orbital surveying of Earth.

Perhaps the greatest achievement of the Gemini series came about by accident. Gemini 6, commanded by former Mercury astronaut Walter Schirra, was intended to catch up and dock with an Agena target vehicle. But the Agena's launch was unsuccessful, and the Gemini 6 flight was postponed. In a brilliant piece of improvisation, it was decided to make Gemini 7 the target for rendezvous by Gemini 6.

In only eight days after the launch of Gemini 7, Gemini 6 had been set up on its rocket on the same pad and readied for launch. Then came a heart-stopping moment. The Titan rocket carrying Schirra and co-pilot Thomas P. Stafford (born in 1930) ignited on schedule–and then shut down. Schirra and Stafford could have ejected from their spacecraft, ruining all hope of a successful mission. They courageously sat tight, and, fortunately, the rocket did not explode. To date, it was the closest the US had come to a manned launch disaster.

Gemini 6 was eventually launched three days later, and performed a brilliantly successful series of manoeuvres in orbit with Gemini 7. At their closest the two craft came within a metre (just over a yard) of each other. After spending a day flying in formation with its sister craft, Gemini 6 returned to Earth leaving Frank Borman and James A. Lovell (both born in 1928) in Gemini 7 to complete a marathon 14-day, 206-orbit flight–a space endurance record that stood until the flight of the Soviet Soyuz 9, four and a half years later.

THE FIRST SPACE DOCKING

It fell to Gemini 8 in March 1966 to make the first space docking. Its commander was to become famous as the first man on the Moon–Neil A. Armstrong (born in 1930).

Armstrong gently nudged the nose of his Gemini spacecraft into the docking collar on the end of the Agena target vehicle, which this time had been launched successfully. But, shortly after docking, trouble hit Gemini 8. A thruster jammed, sending the spacecraft into a rapid and dangerous spin. Armstrong hastily undocked and wrestled to bring the Gemini under control, using up most of the spacecraft's manoeuvring fuel in the process. Ground control ordered the astronauts to return to Earth immediately.

On the next Gemini 9 flight, more trouble occurred when a shroud covering the docking target vehicle failed to eject properly, giving it the appearance of an 'angry alligator' as astronaut Tom Stafford called it. Although the planned docking had to be cancelled, co-pilot Eugene A. Cernan (born in 1934) made a space walk lasting over two

Below: Gemini 6 was launched in December 1965 for a rendezvous with Gemini 7, already in orbit.
Right: Gemini 7 photographed in space through the window of Gemini 6 as the two spacecraft flew close together.

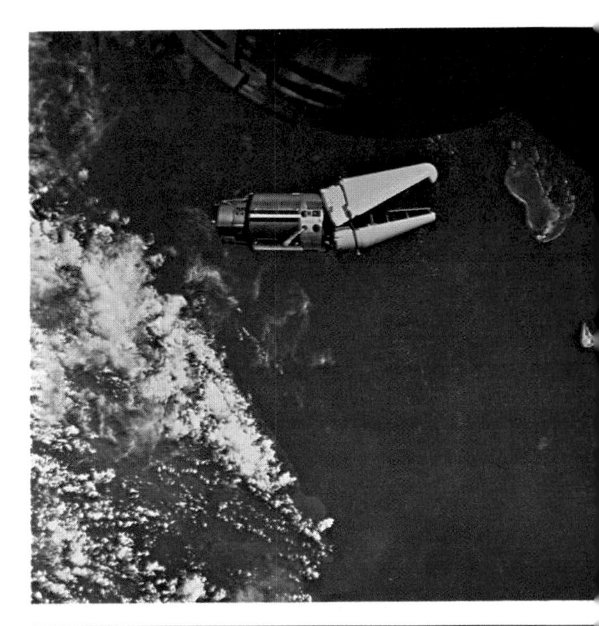

Above: Astronauts David Scott and Neil Armstrong await recovery in their Gemini 8 Capsule after emergency splashdown.
Top right: The shroud of this docking target failed to separate properly, giving it the appearance of an 'angry alligator', as Gemini 9 astronaut Tom Stafford described it.
Right: Gemini 10 approaches an Agena rocket prior to docking with it.
Below right: Gemini 11 looks down from high on Western Australia.

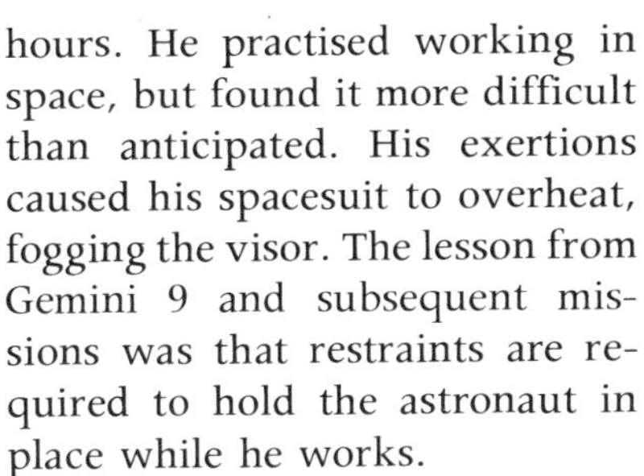

hours. He practised working in space, but found it more difficult than anticipated. His exertions caused his spacesuit to overheat, fogging the visor. The lesson from Gemini 9 and subsequent missions was that restraints are required to hold the astronaut in place while he works.

The three remaining Gemini missions went off almost without hitch. Gemini 10, piloted by John W. Young (born in 1930), docked with an Agena which boosted the astronauts to a higher orbit where they rendezvoused with the Agena target vehicle abandoned by Gemini 8. Co-pilot Michael Collins (born in 1930) made a space walk to this Agena to retrieve a small detector on its outside that measured impacts from tiny meteorites.

Gemini 11's astronauts conducted further rendezvous and docking experiments, and received a spectacular view of Earth when their Agena target vehicle boosted them to a record altitude of 1370 km (850 miles).

Gemini 12 rounded off the series in fine style as astronaut Edwin E. Aldrin (born in 1930) overcame the problems of working in space.

He had been trained in an underwater tank to simulate, as far as possible, the conditions of weightlessness. Special sets of handrails, foot restraints and tethers were provided to help him while he practised cutting wire, turning bolts with a wrench, and using other space tools.

CLOSE-UP OF THE MOON

While this manned space activity was going on, unmanned probes were subjecting the Moon to increasingly close scrutiny. If man was to land there, scientists had to know how safe the lunar surface was. And the necessary information could be obtained only on the spot. In 1964 and 1965, three American Ranger probes sent back an extensive series of photographs as they hurtled into the lunar surface. The Ranger pictures revealed details down to a metre (just over a yard) or so across, far smaller than could be seen through the largest telescope on Earth. Even the apparently flattest parts of the Moon were found to be scattered with small craters and boulders.

In February 1966, mankind received its first photographs from the lunar surface when the Soviet Luna 9 probe landed in the western part of the lowland plain called Oceanus Procellarum. Luna 9 transmitted a panorama of its surroundings, showing the Moon's surface to be granular, like soil, and dotted with small rocks. The myth that the Moon's plains were covered in deep, soft drifts of dust had been finally disproved.

The United States had two Moon-probe programmes designed to return specific information for the Apollo landings. First of these were the Surveyor soft landers, three-legged craft which touched down at walking pace using a system of retro-rockets, similar to that intended for manned landings. By contrast, the landing craft in the Soviet Luna series were reinforced balls which bounced on to the surface of the Moon at high speeds–not a system to be recommended for manned vehicles.

Out of seven attempted Surveyor landings five were successful, and showed that the Moon was perfectly safe for men to land and walk upon. Each Surveyor carried a TV camera which photographed its surroundings in detail. Several of the series carried extra instruments, such as a small box to analyze the composition of the lunar surface and a mechanical arm to dig the soil. The topmost layer of the Moon was found to have been churned to dust by the constant impact of micro-meteorites over many aeons, and the soil composition turned out to be similar to that of volcanic rocks on Earth.

Concurrent with the Surveyor programme were five Lunar

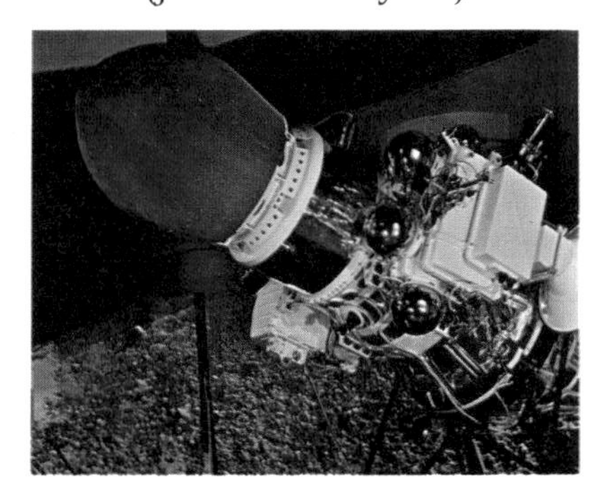

Left: Soviet probe Luna 9 sent the first pictures from upon the surface of the Moon. *Below:* Moon's far side photographed from Apollo 11. *Below right:* Site where Apollo 17 astronauts set up soil-sampling experiments.

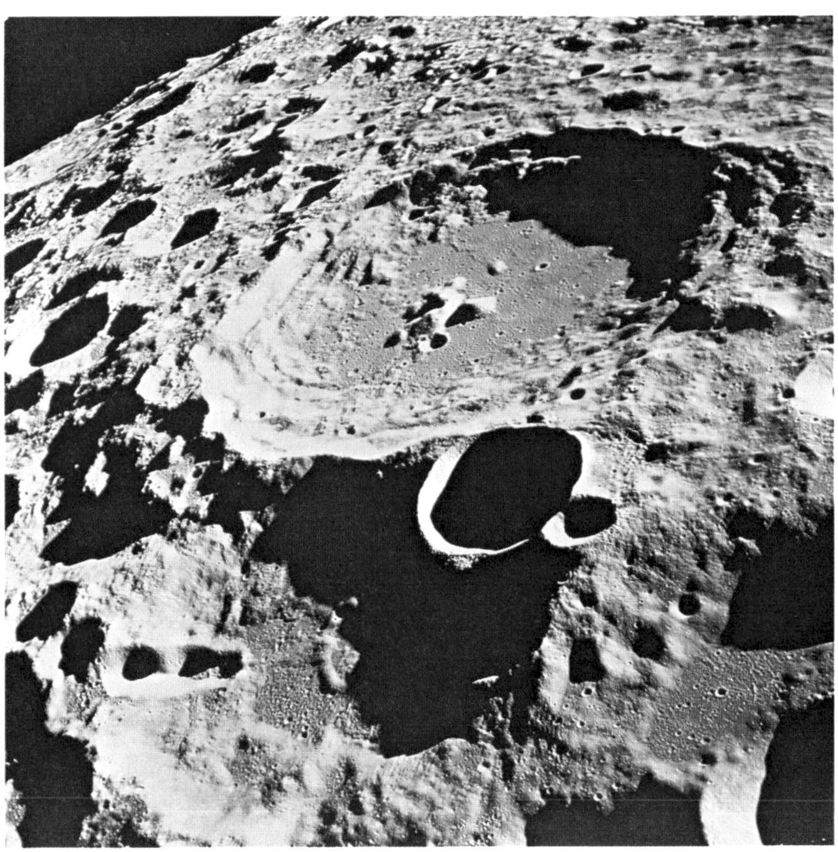

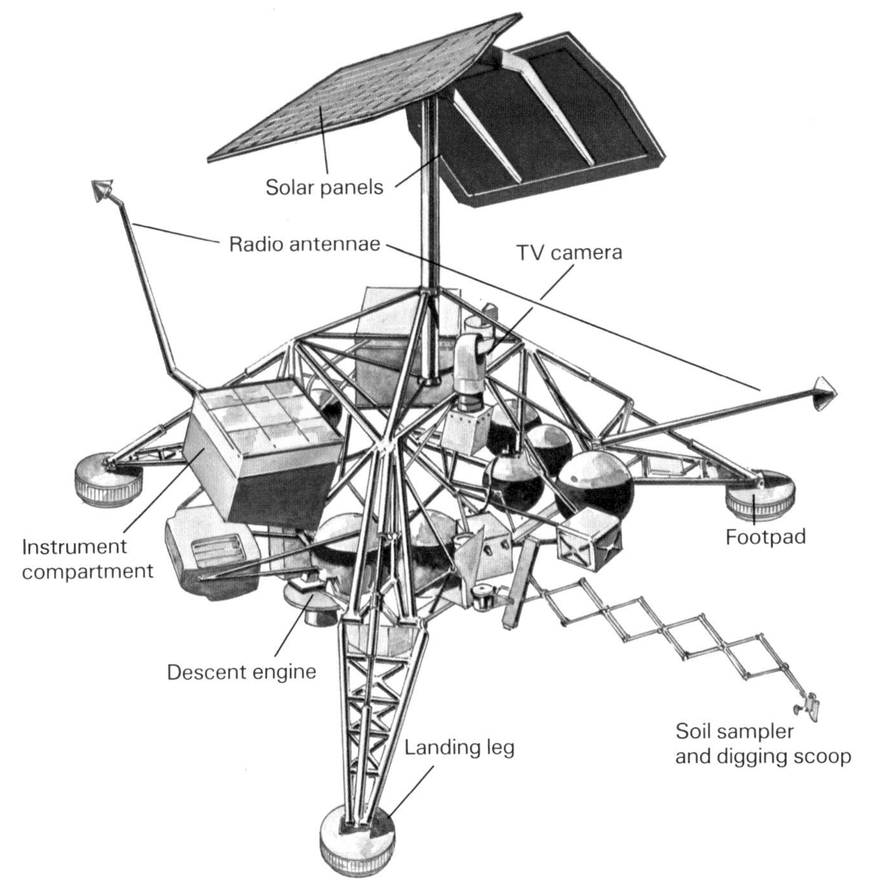

Left: A Surveyor lunar soft lander. *Above:* The charred Apollo capsule in which three astronauts died in a fire during a practice countdown.

Orbiter probes which, as their name implies, scanned the Moon from close orbit. Lunar Orbiter photographs aided the choice of the smoothest and safest landing sites for men, as well as providing astronomers with their most detailed photographic map of the entire Moon.

Tragedy . . .

All seemed set for the first Apollo flights when tragedy struck–not in space but on the ground. To meet the deadline set by President Kennedy, the three-man Apollo spacecraft had been hurriedly prepared for its maiden flight in February 1967. The crew consisted of experienced astronauts Virgil Grissom and Edward White, with novice Roger B. Chaffee. During a practice countdown on January 27, a fire broke out in the spacecraft, apparently due to faulty electrical wiring, and the three men burned to death. The world was horrified.

An immediate investigation recommended numerous improvements to the spacecraft, among them the removal of inflammable materials and an improved hatch design that would allow rapid escape in an emergency. This redesign of Apollo set back the man-on-the-Moon programme by 18 months, and appeared to jeopardize hopes of meeting President Kennedy's deadline.

SOYUZ

While the United States recovered from the blow of the Apollo disaster, the Soviet Union stepped in with a new manned spaceship of its own, called Soyuz. But fate was cruel to both sides in the space race, for the maiden flight of Soyuz was marred by the tragic death of its pilot, Vladimir Komarov (1927–1967).

Soyuz, it subsequently emerged, was a three-part spacecraft for multi-man missions, capable of rendezvous and docking manoeuvres. It was somewhat larger than Gemini but nowhere near the size of Apollo. At the front of Soyuz was a ball-shaped compartment which gave the cosmonauts additional work room while in orbit. The central compartment in which the crew sat was bullet-shaped, 2.2 m (7.2 ft) across its heat shield. This diameter was limited by the width of the launch rocket which was based on the standard Soviet missile used for Sputnik, Vostok and Voskhod launches.

Behind the crew compartment was a cylindrical service module containing air supplies, manoeuvring rockets and solar panels to provide power. The spacecraft's overall weight was 6.7 tonnes. Only the central crew compartment of Soyuz returned to Earth, the other two sections being jettisoned to burn up in the atmosphere. At launch, an escape tower is situated over the spacecraft to pull it to safety in case of a rocket failure.

Soyuz was intended to hold three cosmonauts (later reduced to two for safety reasons), but on its maiden flight it was occupied by only Komarov, commander of the Voskhod 1 mission two and a half years before. Even today, the full story of what went wrong

spinning spacecraft became twisted and the capsule crashed to the ground. Soyuz 1 had made history, but not in the way its makers had intended.

Komarov's death underlined the dangers that still lurked behind every space mission. The designers of Soyuz, like those of Apollo before them, returned to the drawing board to make major modifications.

Soyuz did not re-emerge for 18 months, until October 1968, when Soyuz 3, with one man aboard, rendezvoused in orbit with the unmanned Soyuz 2 but did not dock. But by then its main impact had been dulled by the success of the first Apollo flight, and the Apollo programme overshadowed subsequent Soyuz missions.

Soyuz 4 and 5, both manned, docked with each other and crew members donned spacesuits to walk in space from one craft to the other. Soyuz 6, 7 and 8 made the world's first three-spacecraft flight, though no dockings took place. Soyuz 9 made a record 17-and-a-half-day flight with two men aboard. Later Soyuz missions ferried crews up to Salyut space stations (see Chapter 7).

There is still some doubt about the true purpose of the Soyuz programme. Although Soyuz served very well as an Earth-orbital spacecraft, given a powerful enough booster such as the Proton rocket, it could have taken a man to the Moon. Space experts in the West theorize that just such a project was planned.

Zond

Support for this theory comes from a mysterious series of lunar probes called Zond, launched by Proton rockets, which looped around the Moon before returning to Earth. Pictures show that Zond was apparently a Soyuz without the spherical orbital com-

Above: Soyuz 14 on the launch pad at the Soviet cosmodrome near the town of Tyuratam, north-east of the Aral Sea.
Inset: At the top of the Soyuz spacecraft is a system of escape rockets which would pull the Soyuz to safety in the event of a launch emergency.
Below: Inside Soyuz a cosmonaut takes refreshment.

with Soyuz 1 has not been revealed, but apparently Komarov experienced difficulty in keeping the spacecraft correctly aligned in orbit. After a day in space, he made an emergency re-entry, but the parachute lines of the giddily-

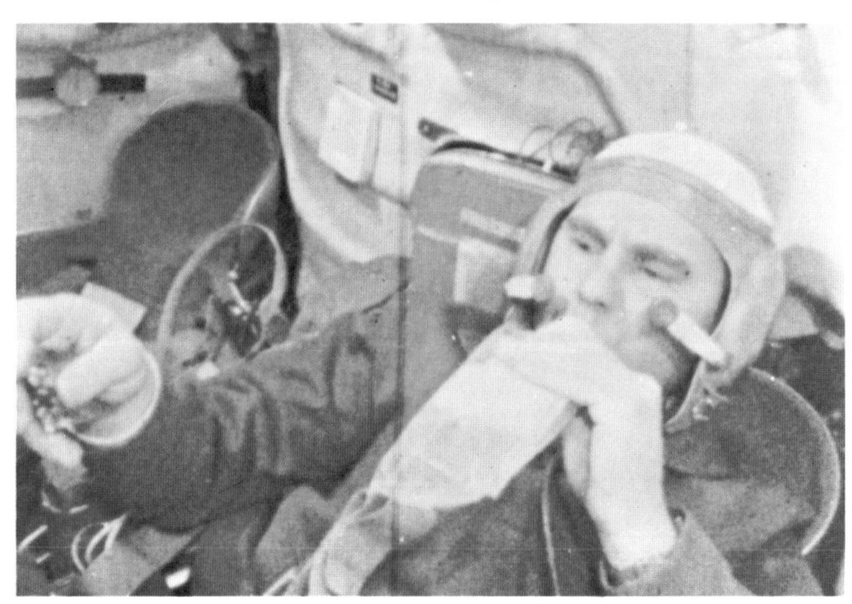

partment at the front. Zond flights around the Moon carried biological specimens to assess radiation hazards, and even transmitted tape recordings of a cosmonaut's voice. These could well have been unmanned test flights of a circumlunar (around the Moon) Soyuz. However when, in December 1968, Apollo 8 sent the first men to orbit the Moon, Soviet interest in Moon flights by Soyuz seemed to vanish.

THE APOLLO MOON-SHIP

Apollo recovered after its fire disaster with an 11-day Earth-orbital mission by Apollo 7 in October 1968. Commander of the crew was Walter Schirra, who thus became the only astronaut to fly all three types of American manned spacecraft.

Apollo's crew compartment, known as the command module, was conical, 3.9 m (12.7 ft) wide at the base and 3.8 m (12.3 ft) high. At its nose was a docking tunnel through which astronauts could crawl into other spacecraft. Behind the conical command module was the service module, a cylinder 7.5 m (24.4 ft) long containing air, water, fuel and power supplies, plus a large engine called the service propulsion system (SPS) for making course changes and entering and leaving lunar orbit. Also attached to the service module were thrusters for use in rendezvous and docking. Total weight at launch was approximately 29 tonnes.

Atop its Saturn launch rocket the Apollo spacecraft was surmounted by an escape tower which would pull it clear in case of emergencies; once the Saturn rocket was safely in flight, this tower was jettisoned. To land on the Moon, a companion spacecraft was needed–the lunar module. This spidery, four-legged craft was tucked below the command

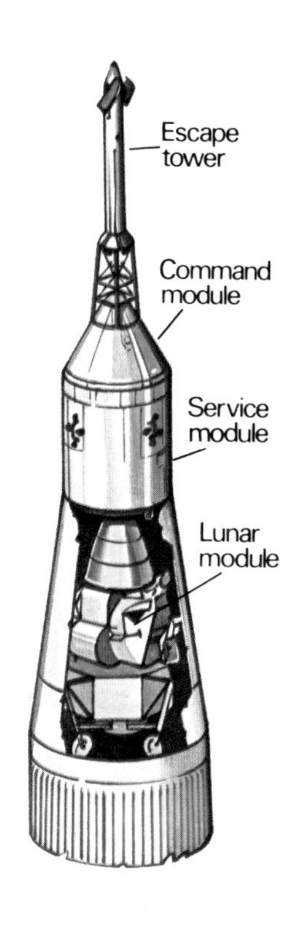

Above: A cutaway diagram of the Apollo assembly with escape tower.
Below: Apollo command and service modules.

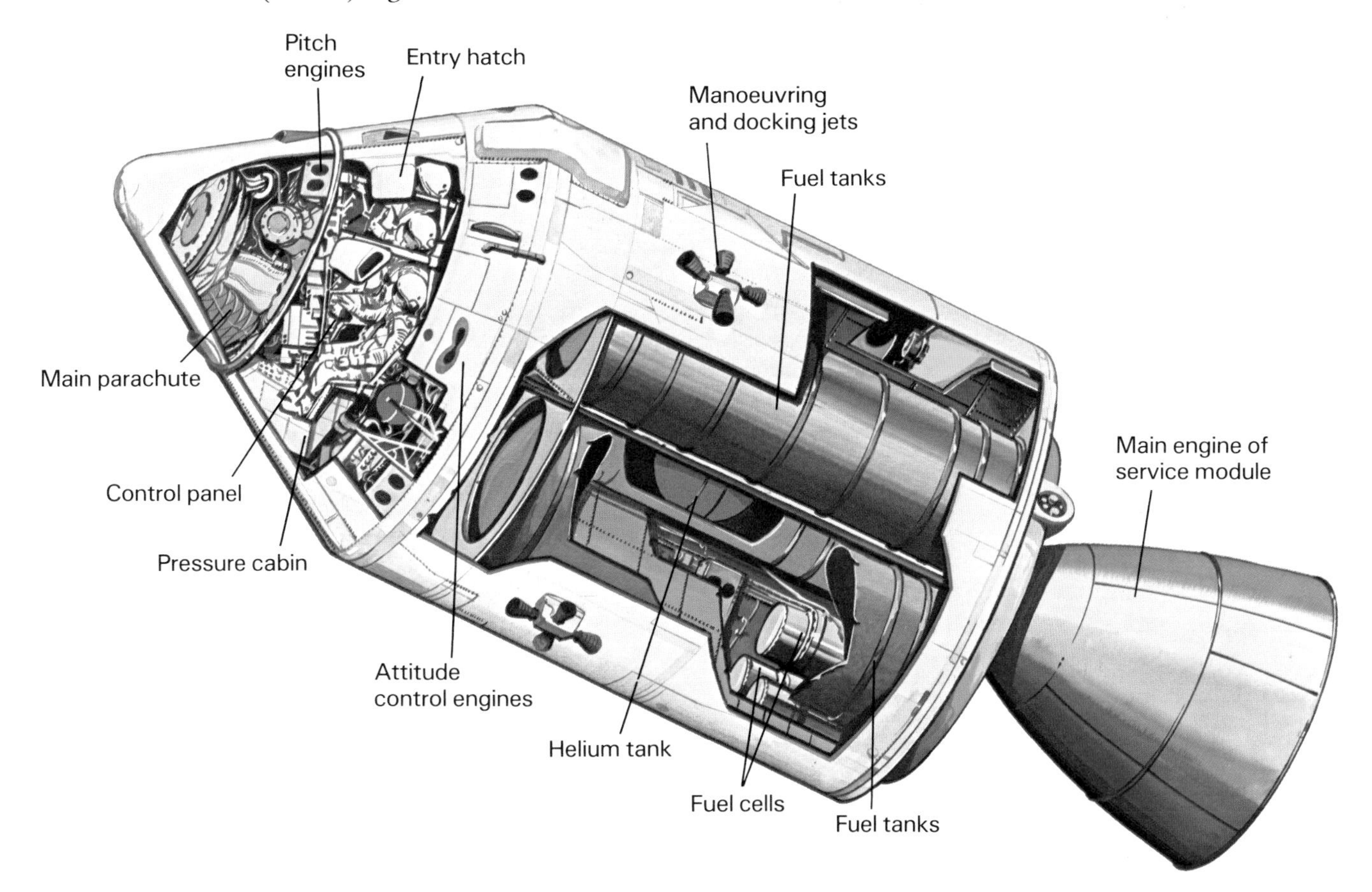

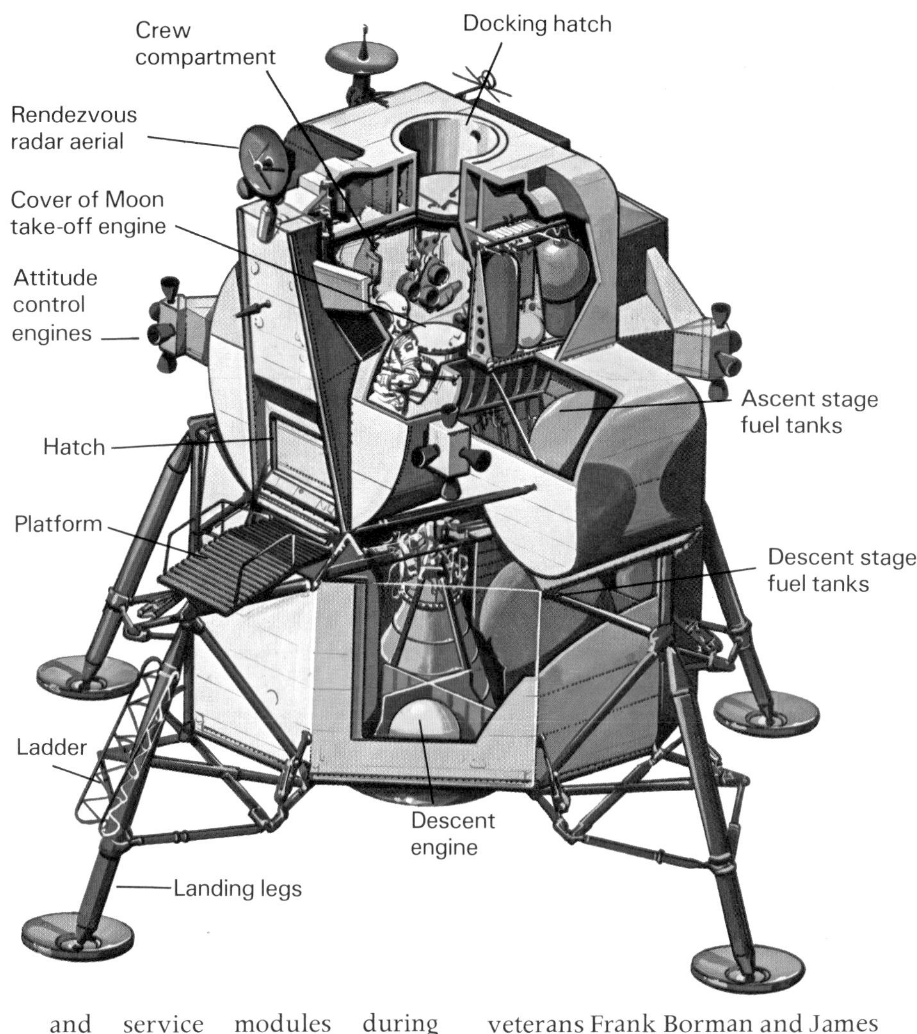

Below: the first stage of a Saturn V rocket is readied for a test firing at the Marshall Space Flight Center, Huntsville, Alabama. *Left:* Cutaway diagram of the lunar module in which astronauts landed on the Moon.

and service modules during launch, and was extracted in space.

On the first two manned Apollo flights the lunar module was not needed. Apollo 7 was a test flight in Earth orbit for the duration of a Moon mission; it included several firings of the big SPS engine that would be needed to place Apollo in lunar orbit and kick it safely back to Earth again. Apollo 7 was put into orbit by the smaller of the two Saturn rockets, Saturn IB. But its successor, Apollo 8, was an altogether more adventurous flight to orbit the Moon, requiring the power of the much larger Saturn V–the first manned launch by the world's greatest rocket.

Risky Mission

Apollo 8 was crewed by Gemini veterans Frank Borman and James Lovell, with novice astronaut William A. Anders (born in 1933). They fully realized that their mission was probably more risky than any before, for they would be unable to return to Earth at a moment's notice. But to keep Apollo on schedule, and to prevent any last-minute Soviet challenge, their mission was vital.

Apollo 8 took off on December 21, 1968, into a so-called parking orbit around Earth, which allowed them to check that all was well before proceeding. Then the third stage of their Saturn V rocket fired to place them on course for the Moon.

Had anything gone wrong on the outward journey, their trajectory allowed them to swing around the far side of the Moon and return safely to Earth. In the event, all went well and the SPS engine fired to place them in orbit a mere 110 km (68 miles) above the Moon's scarred surface. This made them the first humans to come under the gravitational control of a celestial body other than Earth. In awe, the astronauts described their view of both sides of the Moon, and took extensive photographs of possible landing sites.

'The Earth looks pretty small from here,' commented Lovell.

After 10 orbits of the Moon came the crucial re-firing of the SPS engine to put them on course for home; a failure here would have left them stranded in lunar orbit to die. Fortunately the engine proved reliable, as it was to do throughout the Apollo series and, at the end of 1968, the first men to reach the Moon returned safely to a hero's welcome.

One crucial part of Apollo remained to be flight-tested: the lunar module. This was a two-stage craft to enable two astronauts to land on the Moon and take off again. Overall, the lunar module was surprisingly large, standing 7 m (23 ft) high and weighing 15 tonnes at launch. Its lower half, with four landing legs, contained the main engine which

Apollo Flight Plan
The Apollo flight plan required a spacecraft made up of three sections, or modules. The three-man crew were housed in the command module (CM), linked to the main equipment section, the service module (SM). For the outward flight the CM was mated to the lunar module (LM) as shown below. When the craft reached lunar orbit, two of the astronauts transferred to the LM, separated from the CM, and descended to the lunar surface. After their 'moonwalk' they returned to dock with the CM orbiting above. The LM was then discarded before the flight home.

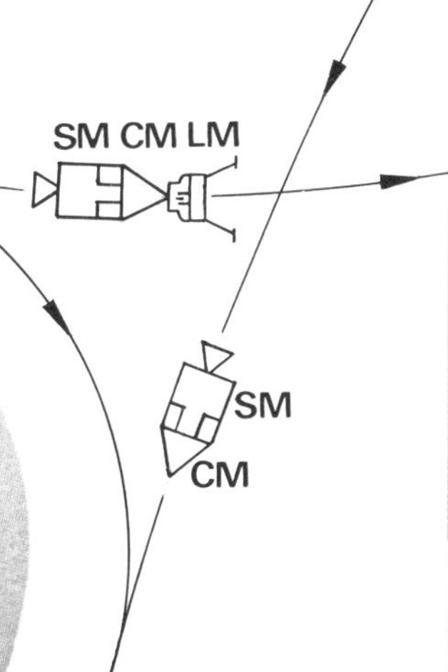

decelerated the craft to a soft touchdown on the Moon.

Once on the Moon, this lower stage served as a launch pad for the upper half containing the astronauts. After their exploration of the lunar surface the astronauts would return to the lunar module and blast off to rendezvous with the command module orbiting above.

Rehearsals

Apollo 9 was to put the lunar module through its paces in the safety of Earth's orbit. First, the lunar module had to be extracted from the third stage of the Saturn V launch rocket by turning the Apollo spacecraft to dock with it. When safely docked, and the unwanted third stage discarded, two men crawled through the docking tunnel into the lunar module. During a brief spacewalk, one astronaut tested a spacesuit designed for use on the Moon. Later, the lunar module was separated from the mother ship as though on a trip down to the Moon's surface, then returned to the command module as though ascending from the Moon's surface. It was a difficult, crucial mission, but it passed without a hitch.

One further step remained before a manned lunar landing: a dress rehearsal in lunar orbit itself. Apollo 10 combined the experience of its two predecessors to put the lunar module through its paces around the Moon, dropping to within 15 km (9.3 miles) of the surface before astronauts Stafford and Cernan piloted the top stage of the lunar module back to command module.

All was now set for the final fulfilment of President Kennedy's daring challenge. Apollo 11 would land men on the Moon.

The Apollo 11 crew: Neil Armstrong, Michael Collins and Edwin Aldrin.

THE FIRST MOON LANDING

Never before or since has so much attention been focused on a space mission. Apollo 11 was scheduled to fulfil an age-old dream of mankind. July 16, 1969, was the fateful day when the Saturn V rocket carrying Apollo 11 rose gracefully into the air from Cape Canaveral. Its crew, whose names were to go down in history, were all experienced astronauts from the Gemini programme: Neil Armstrong, Michael Collins and Edwin Aldrin.

Their launch and outward voyage to the Moon proved faultless, a good omen for their later success. Once safely in lunar orbit, all was prepared for the great moment. Armstrong and

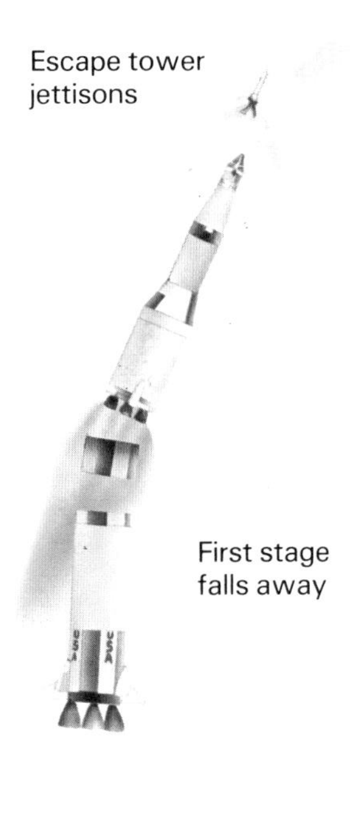

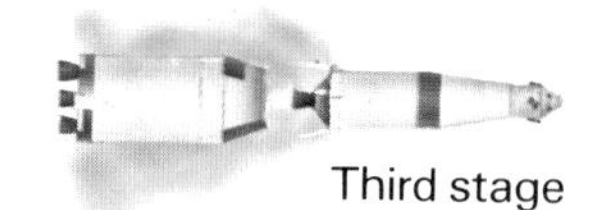

Aldrin crawled into the lunar module, nicknamed Eagle, and separated it for the descent to the selected landing spot in the Sea of Tranquillity, one of the Moon's flat lowland plains.

'The Eagle has wings,' radioed Armstrong.

Firing the braking engine of the lunar module, Armstrong and Aldrin began their descent, monitored by Mission Control in Houston, Texas. Aldrin called out height, speed, and fuel readings as the spidery lunar module dropped ever closer to the surface under guidance from its on-board computer.

Armstrong took over control for the final touchdown as it became apparent that the computer was guiding the craft into a field of boulders. As the world held its breath, Aldrin reported that the Eagle's descent engine was blowing dust around on the

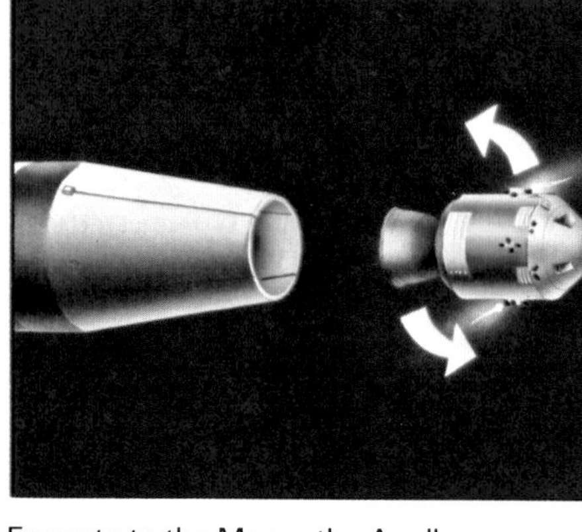

En route to the Moon, the Apollo command and service modules turn to dock with the lunar module which has been stored behind them in the top of the Saturn V's third stage

Bottom left: Launch Control.
Bottom centre: Saturn V, carrying Apollo 11, blasts off.
Far right: Edwin Aldrin steps from the lunar module on to the surface of the Moon.
Centre right: The service and command modules turn to dock with lunar module.

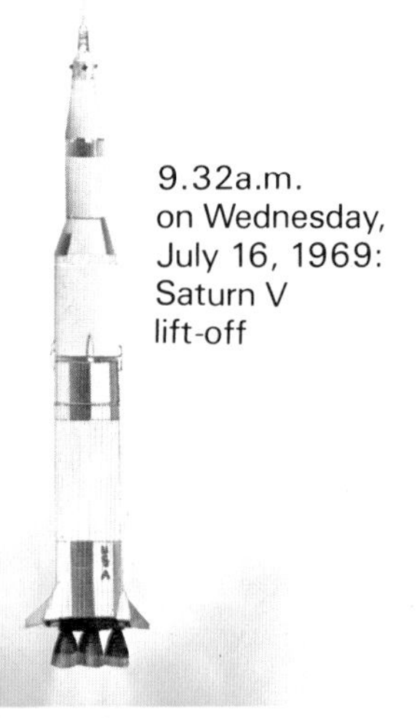

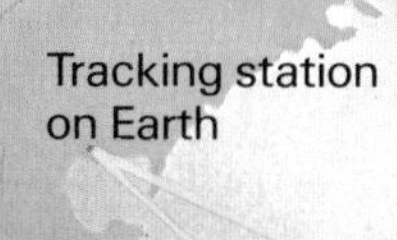

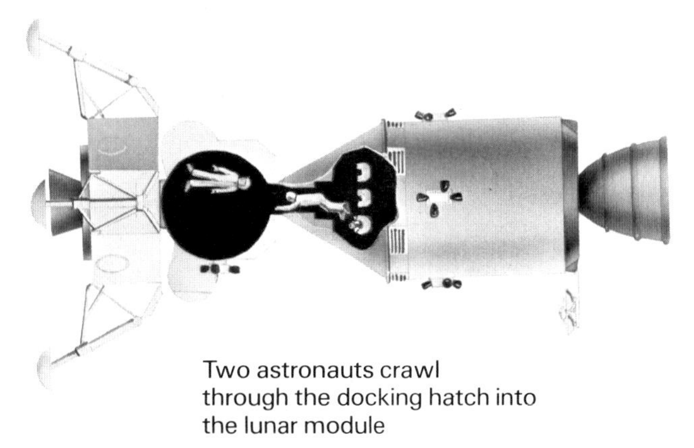

Two astronauts crawl through the docking hatch into the lunar module

lunar surface.

Then came touchdown.

Armstrong radioed: 'Tranquillity Base here. The Eagle has landed.'

At Mission Control, as all around the world, the tension was broken by feelings of relief and delight. It was the evening of July 20, 1969, Greenwich time.

Man on the Moon

Excitedly, Armstrong and Aldrin prepared for their historic first steps on the Moon. Armstrong was first to descend the ladder, activating a small TV camera that monitored his movements.

Command and service modules in lunar orbit

On the Moon. *From left, below:* Edwin Aldrin is photographed on the lunar surface, examining a footpad of the lunar module, and setting out a seismometer to measure Moonquakes with the lunar module seen behind him.
Background picture: The Moon's surface and Earth seen from orbit around the Moon.

Placing his foot on the lunar soil, he said: 'That's one small step for a man–one giant leap for mankind.' He described the lunar soil as like a fine powder, and noted that his boots left distinct imprints. Those footprints will remain on the lunar soil for millions of years, unless destroyed b careless tourists.

Aldrin then descended to joi Armstrong on the Moor Together the men set up equip ment, including a seismometer fc measuring Moonquakes and device to reflect laser beams bac

to Earth. This device has allowed scientists to measure, with astounding accuracy, the distance of the Moon from Earth to within a few centimetres (or inches).

Aldrin practised moving around on the Moon, finding no difficulty despite the low gravity and the slightly slippery nature of the fine soil. Armstrong's Moon walk lasted two and a quarter hours, Aldrin's slightly less. During that time, the two men collected over 20 kg (44 lbs) of rock and soil.

Less than 22 hours after their

Below left: The plaque containing the signatures of Apollo 11 astronauts Armstrong, Collins, and Aldrin, plus that of President Richard M. Nixon, which was attached to a leg of the lunar modul
Below: Aldrin puts out a sail-like device to collect atomic particles from the Sun.

Below, left to right: Lift off from the Moon and return to Earth for the Apollo missions.

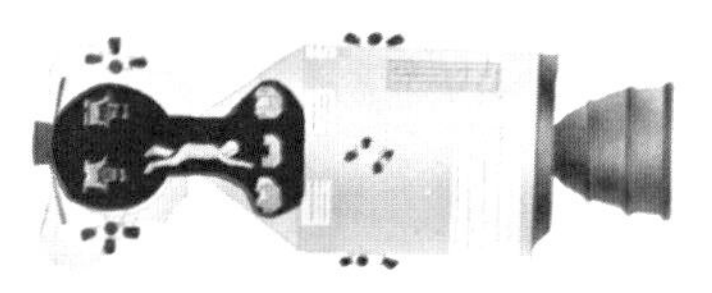

Astronauts transfer into command module for return to Earth

Command module separates from service module for re-entry into Earth's atmosphere

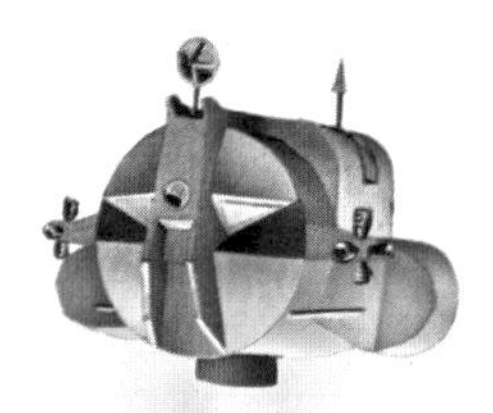

Top half of lunar module lifts off from Moon

landing, Armstrong and Aldrin blasted off from the surface of the Moon in the top half of Eagle to rejoin Collins in the command module orbiting overhead.

Apollo 11 and its crew finally splashed down safely, to find themselves in quarantine for three weeks–a precaution to eliminate any danger of contamination from possible organisms brought back from the Moon. But the Moon turned out to be completely sterile, and the quarantine procedures were abolished for later missions.

Lunar Findings

Most astounding of the findings from the Apollo 11 samples was the extreme age of the Moon–3700 million years at the Tranquillity Base site, which subsequent missions showed was one of the youngest parts of the lunar surface. The rocks turned out to be like basalt lava on Earth, indicating that they had once been molten and apparently flowed out on to the surface from the Moon's interior.

For all Apollo 11's success, it was nearly upstaged by a mys-

Left: Armstrong and Aldrin return from the surface of the Moon in the ascent stage of the lunar module.
Above and top: A Moon rock, and colourful minerals in a sample of Moon rock seen through a microscope.

terious Soviet Moon probe called Luna 15 which crashed on the Moon the same day as Apollo 11's successful touchdown. It later became clear that Luna 15 had been an attempt to snatch a small lunar sample.

This was finally achieved the following year by the robot probe Luna 16, which scooped up a mere 100 grams (3.5 oz) of lunar topsoil and automatically brought it back to Earth. But by then both Apollo 11 and Apollo 12 had gathered many kilograms (or pounds) of Moonstuff.

Nevertheless, the automatic return of lunar soil by Luna 16, and later by two more craft, was a major technical achievement, and is a technique that can be of particular use in sampling areas too dangerous for manned landings, such as the lunar far side. Similar craft can also be used to bring soil from Mars or other bodies.

President Nixon welcomes the Apollo 11 astronauts home.

Above: Studying the spoils — a Moon rock in the laboratory.
Right: Quarantine procedures after splashdown: astronauts wear special gear to prevent the spread of germs they might have brought from the Moon.

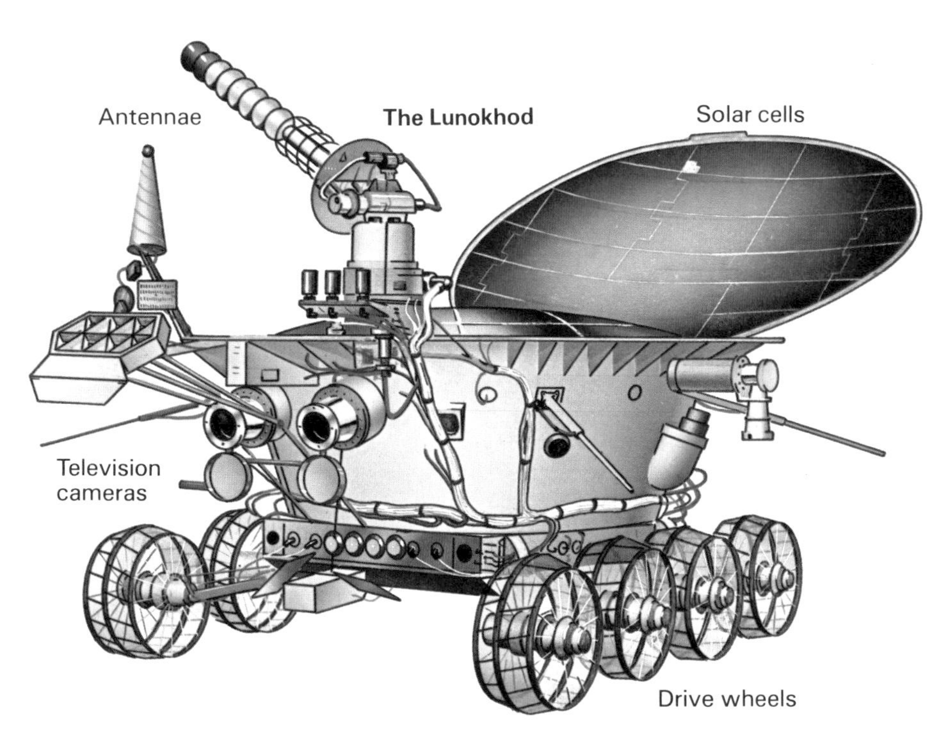

Above: Return to the Moon. The crew of Apollo 12 on their way to the launch pad for the second lunar landing mission.
Left: Lunokhod, a Soviet Moon car, was steered by remote control from Earth.

The Soviets scored another success in their unmanned lunar programme in 1970 with the Lunokhod automatic Moon rover. Lunokhod, and an improved successor in 1973, were driven around by remote control from Earth. Lunokhod's TV 'eyes' allowed ground controllers to survey the lunar landscape. Such craft could be employed for exploring the planets.

A second Apollo mission beat President Kennedy's deadline for landing men on the Moon. Apollo 12 in November 1969, commanded by Charles Conrad, touched down in the Oceanus Procellarum only a few hundred metres (or yards) from the Surveyor 3 robot probe which had landed two and a half years before. Conrad and his partner Alan L. Bean (born in 1932) made two lunar walks, setting up scientific experiments and strolling over to the long-dead Surveyor, from which they retrieved a number of parts, including the TV camera and soil sampling arm, to bring back to Earth for analysis. Scientists found terrestrial micro-organisms on the camera that had apparently survived in the harsh lunar environment.

Apollo 12 returned to Earth with 34 kg (74 lbs) of lunar rocks. These turned out to be 3300

Below: Apollo 12 astronaut Alan Bean approaches the Surveyor 3 robot soft lander.
Below right: Bean puts out scientific equipment on the Moon.

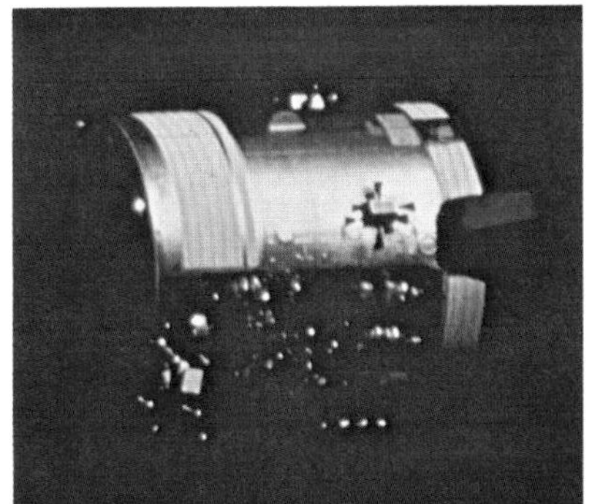

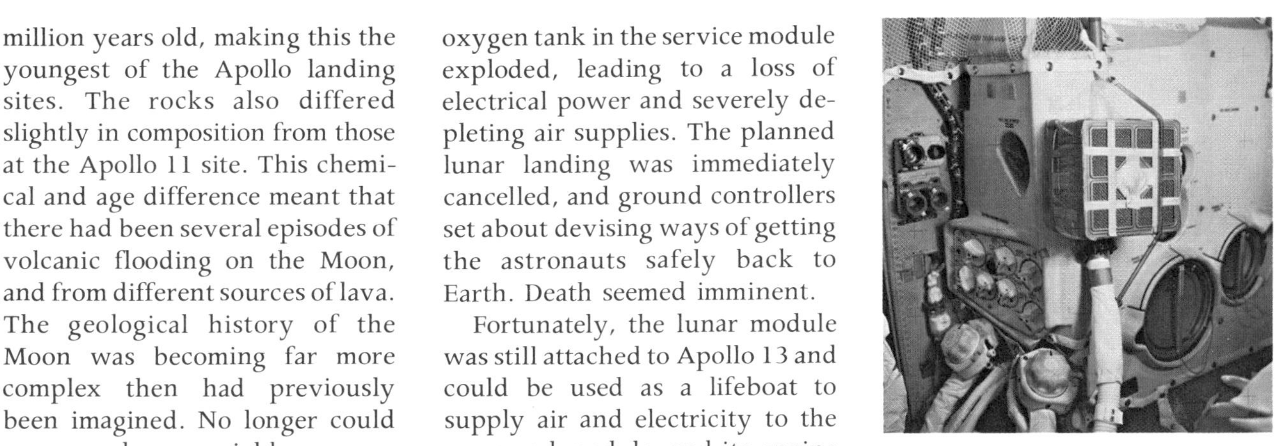

million years old, making this the youngest of the Apollo landing sites. The rocks also differed slightly in composition from those at the Apollo 11 site. This chemical and age difference meant that there had been several episodes of volcanic flooding on the Moon, and from different sources of lava. The geological history of the Moon was becoming far more complex then had previously been imagined. No longer could we regard our neighbour as a lump of cold, dead rock marked only by meteorites. It now appeared as though the Moon also had an internal life of its own.

Unlucky 13

Apollo 13, true to its number, was the unlucky mission in the series. On the way to the Moon, an oxygen tank in the service module exploded, leading to a loss of electrical power and severely depleting air supplies. The planned lunar landing was immediately cancelled, and ground controllers set about devising ways of getting the astronauts safely back to Earth. Death seemed imminent.

Fortunately, the lunar module was still attached to Apollo 13 and could be used as a lifeboat to supply air and electricity to the command module, and its engine could be used for changes of course. The crippled Apollo 13 swung behind the Moon and returned to Earth, becoming uncomfortably cold and stuffy inside. Just before re-entry the life-saving lunar module was jettisoned, and Apollo 13 splashed down safely.

Above left: Apollo 12 astronaut Alan Bean removes scientific equipment from a bay on the lunar module.
Top: The Apollo 12 lunar module descends to the Moon's surface.
Middle: Damage to Apollo 13's service module which had been caused by an oxygen tank explosion en route to the Moon.
Above: The Apollo 13 astronauts rigged up this device in the lunar module to help purify carbon dioxide from the air aboard their crippled spacecraft.

What could have been a horrific space tragedy had become a triumph of improvisation. All future Apollo service modules were modified to prevent a repeat disaster.

Insets:- Below left: Apollo 16 lunar module takes off from the Moon.
Below right: American flag set up by Apollo 17 astronauts.
Bottom left: Apollo 16 lunar module after leaving the Moon's surface.
Bottom right: Apollo 17 command and service modules in orbit.

MOON STUDIES

Alan Shepard, America's first astronaut, returned to space to command the flight of Apollo 14. It was to be the first to visit an upland region of the Moon, in this case the hilly formation called Fra Mauro believed to be composed of rocks thrown out from the impact that carved the Mare Imbrium basin. Among the targets of the Apollo 14 astronauts' two Moon walks was Cone crater, believed to have been formed by a meteorite impact. The rocks at Fra Mauro were much lighter coloured than the dark lavas of the lowlands because of their different composition, and dated back to about 3900 million years ago, which was evidently the time of the Mare Imbrium impact.

For the final three Apollo Moon landings considerable improve-

ments were introduced, notably the inclusion of an electrically-powered lunar car, stored aboard the lunar module, which enabled astronauts to drive about over the surface of the Moon and explore further afield.

Apollo 15 was the first such improved mission, landing near Hadley Rille, a winding valley at the foot of the lunar Apennine mountains. Astronauts David R. Scott (born in 1932) and James B. Irwin (born in 1930) stayed for two and a half days on the Moon, making three exploratory trips and driving over 27 km (17 miles) in the lunar rover. During their excursions, on which they collected 77 kg (170 lbs) of samples, both ground controllers and public followed the astronauts' movements by means of a TV camera aboard the rover, which sent colour pictures to Earth via an on-board antenna.

Scott and Irwin drove to the edge of Hadley Rille, where they saw layers of lava in the rille walls; evidently the rille itself was

Below left (inset): Receiving data on Earth from seismometers set up on the Moon.
Background panorama: Apollo 17 astronaut Jack Schmitt is dwarfed by a huge lunar boulder. The electric powered lunar roving car can be seen to the right of the boulder.

carved out by a later flow of lava. They also photographed what appeared to be layers in the rocks of the nearby Apennine mountains.

Despite expectations that the mission would return with rocks dating back to the origin of the Moon 4600 million years ago, the oldest rocks found were about 4200 million years old. Some were no older than the ones found at the Apollo 14 site.

Searching for Fragments

The search for fragments of the Moon's ancient crust was continued by Apollo 16, which landed in a rugged highland area near the crater Descartes. Geologists expected this area to consist of lava flows far more ancient than those of the lowland plains–but they were wrong. Astronauts John W. Young (born in 1930) and Charles M. Duke (born in 1935) soon realized that the rocks

VOSKHOD FLIGHTS

Mission	*Launch Date*	*Remarks*
Voskhod 1	October 12, 1964	Vladimir Komarov, Konstantin Feoktistov and Boris Yegorov made day-long flight.
Voskhod 2	March 18, 1965	Pavel Belyaev and Alexei Leonov made day-long flight, including spacewalk.

AMERICAN MANNED GEMINI FLIGHTS

Mission	*Launch Date*	*Remarks*
Gemini 3	March 23, 1965	Virgil Grissom and John Young made three-orbit test flight.
Gemini 4	June 3, 1965	James McDivitt and Edward White orbited Earth 62 times. White made space walk.
Gemini 5	August 21, 1965	L. Gordon Cooper and Charles Conrad made 120-orbit, eight-day flight.
Gemini 7	December 4, 1965	Frank Borman and James Lovell made 206-orbit, 14-day flight.
Gemini 6	December 15, 1965	Walter Schirra and Thomas Stafford rendezvoused with Gemini 7 during 15-orbit flight.
Gemini 8	March 16, 1966	Neil Armstrong and David Scott made first space docking.
Gemini 9	June 3, 1966	Thomas Stafford and Eugene Cernan made 45-orbit flight, including rendezvous manoeuvres and space walks.
Gemini 10	July 18, 1966	John Young and Michael Collins made 43-orbit flight, with docking and space walks.
Gemini 11	September 12, 1966	Charles Conrad and Richard Gordon made 44-orbit flight, with docking and space walk.
Gemini 12	November 11, 1966	James Lovell and Edwin Aldrin made 59-orbit flight, including docking and work in space.

around them were broken and shattered ejecta from an ancient impact which had presumably excavated one of the Moon's great lowland basins, possibly Mare Orientale. The Descartes highlands were therefore similar to the Fra Mauro area sampled by Apollo 14. Since the Descartes rocks had been broken up and re-melted, it was difficult to tell how old they were or what their original composition had been.

Home again! Apollo 17 heads for splashdown under three parachutes.

Perhaps Apollo 17 would have more luck. Its target was an area of the eastern side of Mare Serenitatis, between the Taurus mountains and the crater Littrow. Here, dark lava met bright highlands and offered many rock samples. Apollo 17 astronauts Gene Cernan and Harrison H. Schmitt (born in 1935) set new records by exploring the lunar surface for 22 hours and driving 35 km (22 miles) during their 75 hours on the Moon, when they collected 113 kg (250 lbs) of rocks.

AMERICAN MANNED APOLLO FLIGHTS

Mission	*Launch Date*	*Remarks*
Apollo 7	October 11, 1968	Walter Schirra, Donn Eisele and Walter Cunningham made 11-day test flight in Earth orbit.
Apollo 8	December 21, 1968	Frank Borman, James Lovell, and William Anders made 10 orbits of the Moon.
Apollo 9	March 3, 1969	James McDivitt, David Scott and Russell Schweickart tested lunar module in Earth orbit.
Apollo 10	May 18, 1969	Thomas Stafford, John Young and Eugene Cernan made dress-rehearsal for Moon landing.
Apollo 11	July 16, 1969	Neil Armstrong, Michael Collins, and Edwin Aldrin made first Moon-landing flight. Lunar module touched down in Sea of Tranquillity on July 20.
Apollo 12	November 14, 1969	Charles Conrad, Richard Gordon and Alan Bean. Conrad and Bean landed in Ocean of Storms on November 19.
Apollo 13	April 11, 1970	James Lovell, John Swigert and Fred Haise had landing attempt cancelled after explosion in service module.
Apollo 14	January 31, 1971	Alan Shepard, Stuart Roosa, Edgar Mitchell. Shepard and Mitchell landed in Fra Mauro region on February 5.
Apollo 15	July 26, 1971	David Scott, Alfred Worden and James Irwin. Scott and Irwin landed at Hadley Rille on July 30.
Apollo 16	April 16, 1972	John Young, Thomas Mattingly and Charles Duke. Young and Duke landed in Descartes highlands on April 21.
Apollo 17	December 7, 1972	Eugene Cernan, Ronald Evans and Harrison Schmitt. Cernan and Schmitt landed on edge of Sea of Serenity on December 11.

Orange-Coloured Soil

One exciting discovery was orange-coloured soil, which was at first suspected to be due to recent volcanic activity. But geologists on Earth found it to consist of small glass spheres apparently fused by the heat of a meteorite impact; the soil itself was the same age as the rest of the surrounding mare surface–3700 million years, virtually identical with the age of the Apollo 11 site. Careful searching among the Apollo 17 samples eventually revealed two rocks from the lunar highlands that seem to date back 4600 million years, thus providing samples of the original lunar crust.

When Apollo 17 splashed down on December 19, 1972, it brought the first era of manned lunar exploration to a close. There had been plans for further landings, but these were cancelled due to budgetary cuts. Enthusiasm for space had waned since the heady days of the 1960s and, without a realistic Soviet challenge, the political impetus of Apollo was gone. Nonetheless it had been an exciting, impressive achievement.

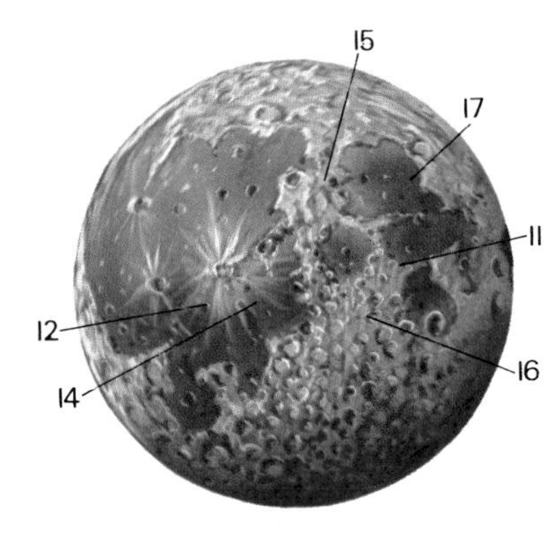

Top: Apollo 17 astronaut Jack Schmitt collects Moon samples with a rake. *Above:* Moon landing sites.

Beyond that, what did Apollo and its 365 kg (805 lbs) of lunar samples tell us about the Moon? Many of the details remain controversial, but it seems clear that, like the rest of the solar system, the Moon formed 4600 million years ago, although whether it formed in orbit around Earth or was an independent body that was later captured is uncertain.

Evidently, the Moon was molten shortly after its birth, and an ancient crust settled around it like scum. Then came an episode of heavy bombardment as the Moon swept up left-over rocks from space. This was the time when the major formations of the lunar surface were blasted out.

Lunar Volcanism

About 3900 million years ago this bombardment waned, to be replaced by a more gentle period of volcanic oozing as molten lava spewed out every now and again over the Moon's surface, filling the lowlands. Inwardly, the Moon is still slightly warm, but the main phase of lunar volcanism seems to have died out over 3000 million years ago. Since then the Moon has changed little, while the much more geologically active Earth has been undergoing vast upheavals that have destroyed its earliest rocks. Therefore, knowledge of the Moon helps us fill in the early stages of Earth's evolution.

Men will eventually return to the Moon, although it is impossible to say when. They will set up bases and perhaps even mine the surface, as speculated in Chapters 7 and 8. Whatever happens, none of it would be possible if it hadn't been for the pioneering explorations of Apollo.

EARLY SOYUZ FLIGHTS

Mission	*Launch Date*	*Results*
Soyuz 1	April 23, 1967	Vladimir Komarov killed on re-entry.
Soyuz 2	October 25, 1968	Unmanned target for Soyuz 3.
Soyuz 3	October 26, 1968	Georgi Beregovoi manouevred close to Soyuz 2 but did not dock.
Soyuz 4	January 14, 1969	Vladimir Shatalov docked with Soyuz 5.
Soyuz 5	January 15, 1969	Boris Volynov, Yevgeny Khrunov, and Alexei Yeliseyev docked with Soyuz 4.
Soyuz 6	October 11, 1969	Georgi Shonin and Valeri Kubasov carried out joint manoeuvres with Soyuz 7 and 8.
Soyuz 7	October 12, 1969	Anatoli Filipchenko, Vladislav Volkov and Viktor Gorbatko carried out manoeuvres with Soyuz 6 and 8.
Soyuz 8	October 13, 1969	Vladimir Shatalov and Alexei Yeliseyev commanded group flight with Soyuz 6 and 7.
Soyuz 9	June 2, 1970	Andrian Nikolayev and Vitaly Sevastyanov made $17\frac{1}{2}$-day flight, then a record.

(Table continued on page 126)

5 SATELLITES FOR MAN

Manned space missions receive most publicity, but they account for only a few per cent of all spacecraft launches. For instance, by the end of 1976 over 2000 satellites and probes had been launched by all nations, but of these only 59 were manned. Of this total, 884 craft remained in space, the rest having re-entered the atmosphere or landed on the Moon or a planet.

The pace of space launchings has increased greatly since the first Sputniks, as shown by the following figures: in 1958, the first full year of space activities, seven spacecraft were launched by the United States and one by the Soviet Union; in 1976, the Soviet Union launched a record total of 99 spacecraft, against 26 by the United States; the US total reached its own peak of 70 successful launches in 1966.

Largely unheralded, a host of satellites are helping mankind in many ways, from communications around the globe to position finding, weather watching, and simply keeping an inventory of the resources of our small planet.

COMMUNICATION SATELLITES

Among the seven American craft launched in that first year of space was a satellite called Score (Signal Communication by Orbiting Relay Equipment). This consisted of an Atlas booster with a tape recorder on board, which received and transmitted voice messages to and from the ground. This was the first, if crude, communications satellite.

The advantage of a satellite for communications is that it can be seen from a very wide area of Earth, like a super-tall broadcasting tower, and it is not affected by the radio blackouts that can interrupt signals bounced off the ionosphere. The higher this 'tower in the sky', the greater the area of Earth it can be seen from.

Communications satellites do away with the need for expensive

Right: Diagram to show NASA's complex communications network.
Below: Early Bird, the first of the Intelsat series of communications satellites. It was launched over the Atlantic in 1965 and linked the US and Europe by telephone and television for 4 years.

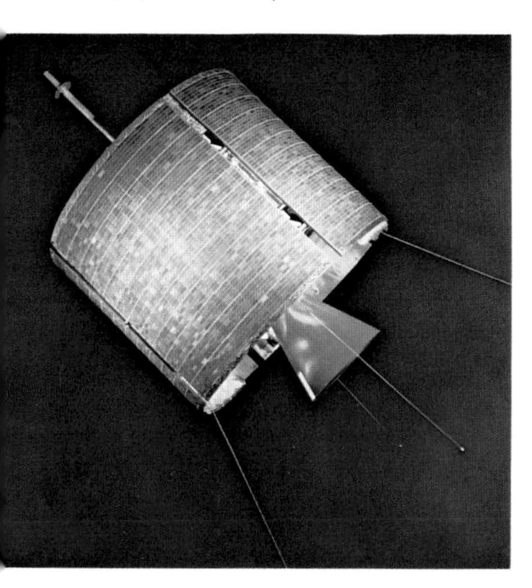

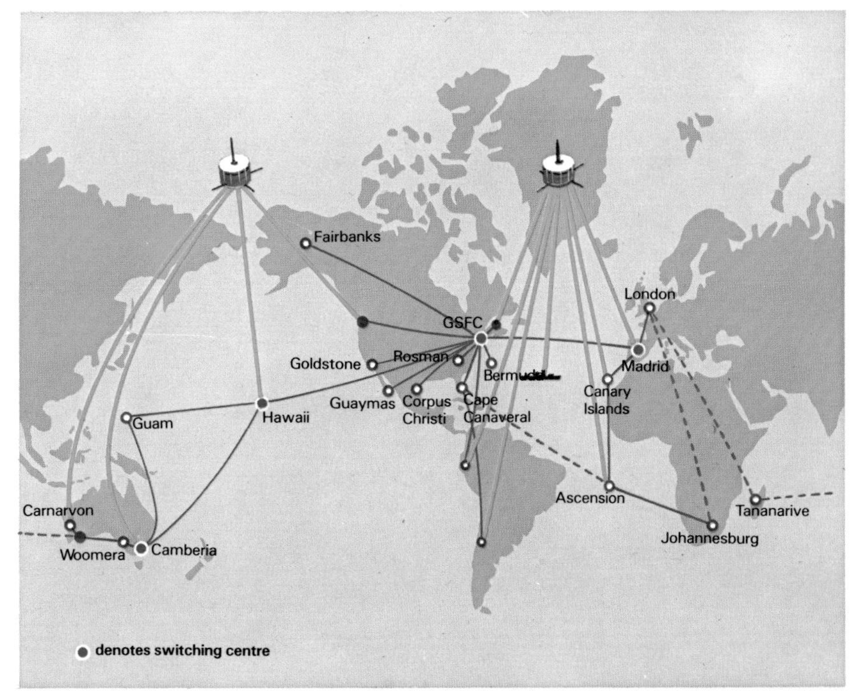

systems of cables linking countries, and they allow transmission of colour TV programmes around the world which would be difficult or impossible to do by cable or chains of relay stations. These satellites have improved world communications and made them relatively cheaper.

Military forces need good communications, and it was the US Army that made the early running in this field. In 1960, it orbited a satellite called Courier which relayed communications and data between military ground stations. Courier was a satellite of the type known as an active repeater, because it received and then retransmitted messages. But experiments were also undertaken by NASA with so-called passive-reflector type satellites – simply balloons with shiny surfaces that reflected radiation beamed at them.

Two such balloon satellites, called Echo, were launched in 1960 and 1964 respectively. Echo 1 was 30 m (98 ft) in diameter, and its successor even larger. They orbited at heights of about 1500 km (930 miles). Their shiny surfaces made them easily visible as they drifted across the sky like bright stars cast adrift. Both eventually re-entered the atmosphere and burned up. Passive-reflector satellites have the advantage that there are no electronic parts to fail, and that anyone can bounce signals off them. But the reflections received on Earth are inevitably very weak, and passive reflectors are no match for communications satellites embodying modern electronic amplifiers.

Telstar–a New Era for Communications

This fact was spectacularly demonstrated in 1962 by Telstar, which sent the first live TV pictures across the Atlantic, thereby opening a whole new era of communications. When not carrying TV, it could handle up to 600 telephone calls. Telstar, and a similar satellite called Relay, demonstrated the value of active-repeater communications satellites.

But Telstar and Relay suffered from one major problem. They were in orbits that took them around Earth every three hours or so, which meant that ground stations had to track them continuously as they moved across the sky, and they were in view from two ground stations at the same time for only a limited period.

One way of setting up permanent communications between two ground stations is to have a succession of such satellites, although this does not eliminate the problem of tracking them as they move. The best answer is to use a much higher orbit in which the satellite seems to remain stationary over Earth.

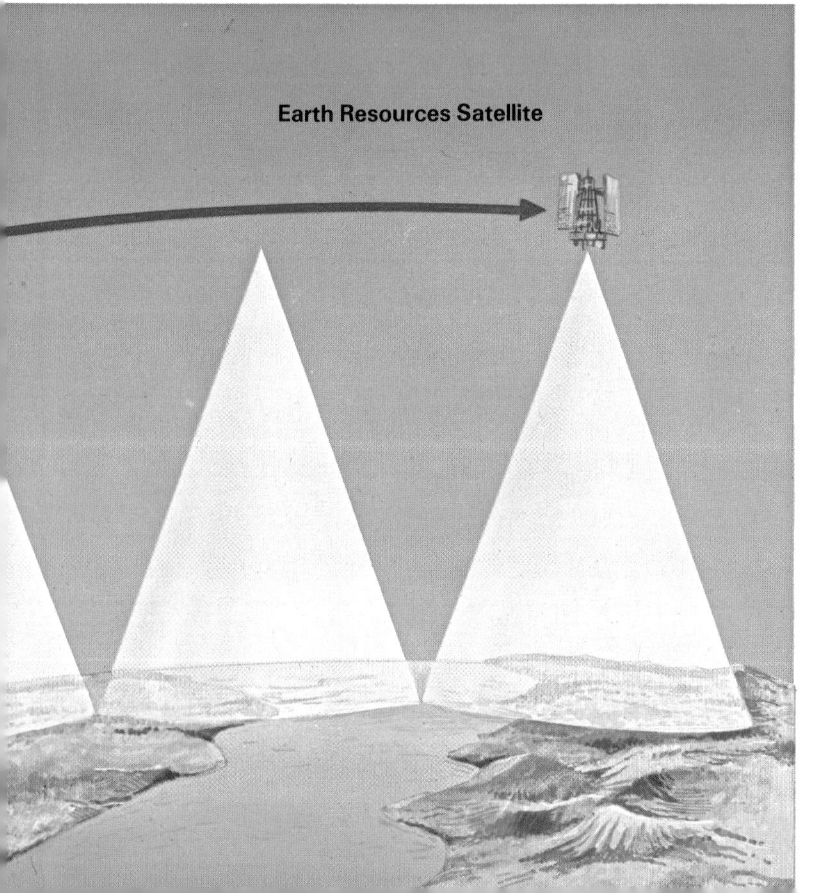

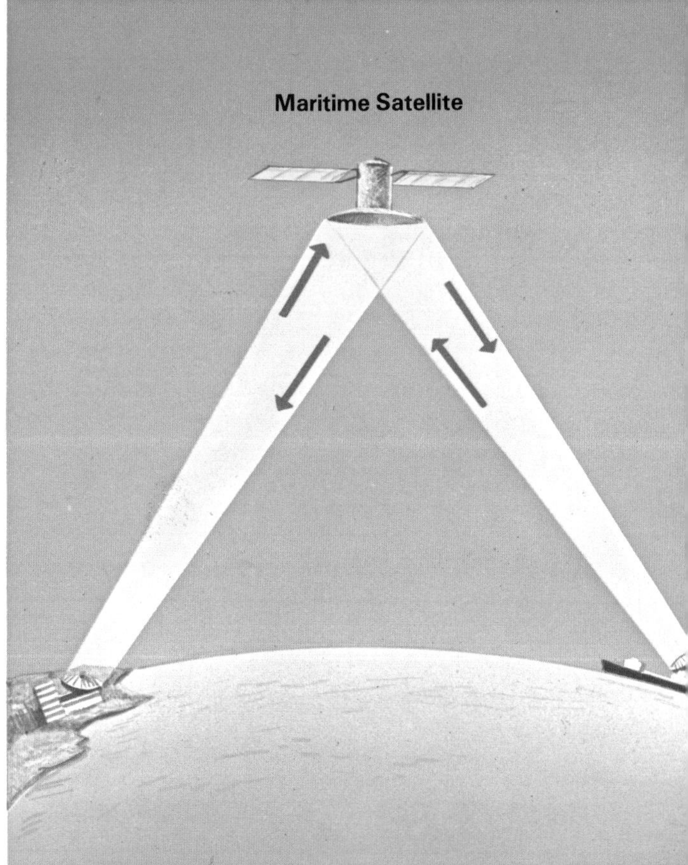

Earth resource satellites photograph Earth in search of natural resources, and help identify crop disease and water pollution.
Maritime satellites are used for ship-to-shore communications and ensure that distress calls don't go unheeded.
Educational satellites. In 1975, NASA's ATS 6 satellite was stationed over India to broadcast education programmes to small ground stations.
Communications satellites. Intelsat IVA is capable of handling as many as 9000 telephone calls.

The Right Orbit

As mentioned in Chapter 3, the further away a satellite is from Earth the slower it need move to stay in orbit. There is a point, 35 900 km (22 300 miles) above Earth, at which a satellite orbits in the same time as it takes Earth to spin on its axis–once every 24 hours. A satellite in such an orbit directly above the equator would therefore appear to hang stationary in the sky, for it would be turning in space at the same rate as the spinning Earth. Such an orbit is known as a geostationary or synchronous orbit.

The ingenious idea of placing a satellite in such an orbit for communications was suggested as long ago as 1945 by the English science-fiction and space writer Arthur C. Clarke (born in 1917), and has since been enthusiastically adopted not just for communications satellites but for many other kinds of satellite as well.

To get into geostationary orbit, a satellite is first of all launched into a highly elliptical orbit, the most distant point from Earth (the apogee) of which is at the required 35 900 km (22 300 miles). At this distance, an extra engine is fired to circularize the satellite's orbit so that it remains a constant 35 900 km (22 300 miles) above the equator. If the satellite's orbit is inclined to the equator, it will appear to move alternately north and south of the equator as it orbits Earth. Therefore one could not, for instance, have a geostationary satellite orbiting permanently above Moscow, London, or Washington.

Left: Telstar, the pioneering communications satellite which sent the first live TV pictures across the Atlantic in 1962.

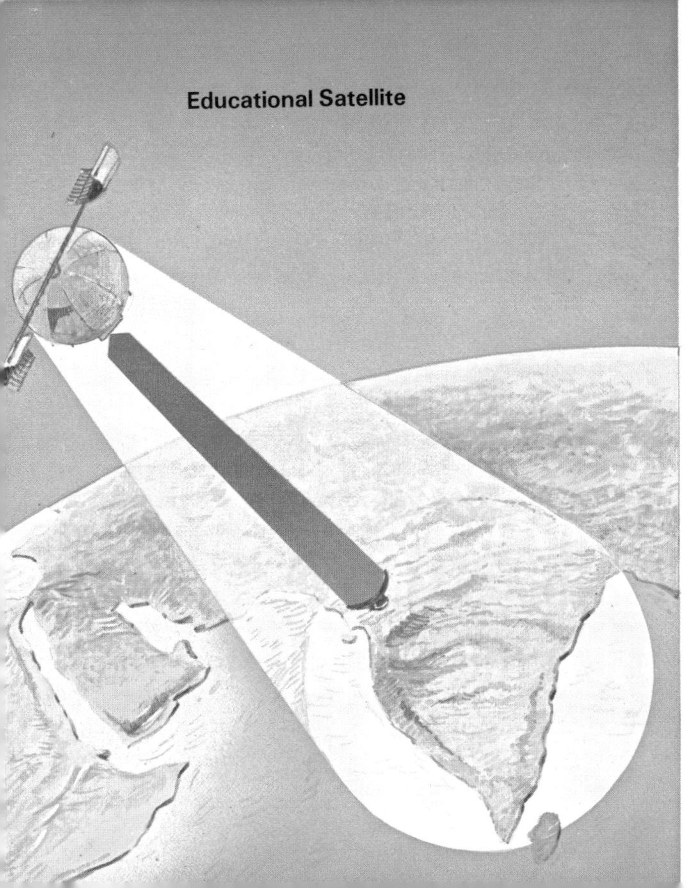

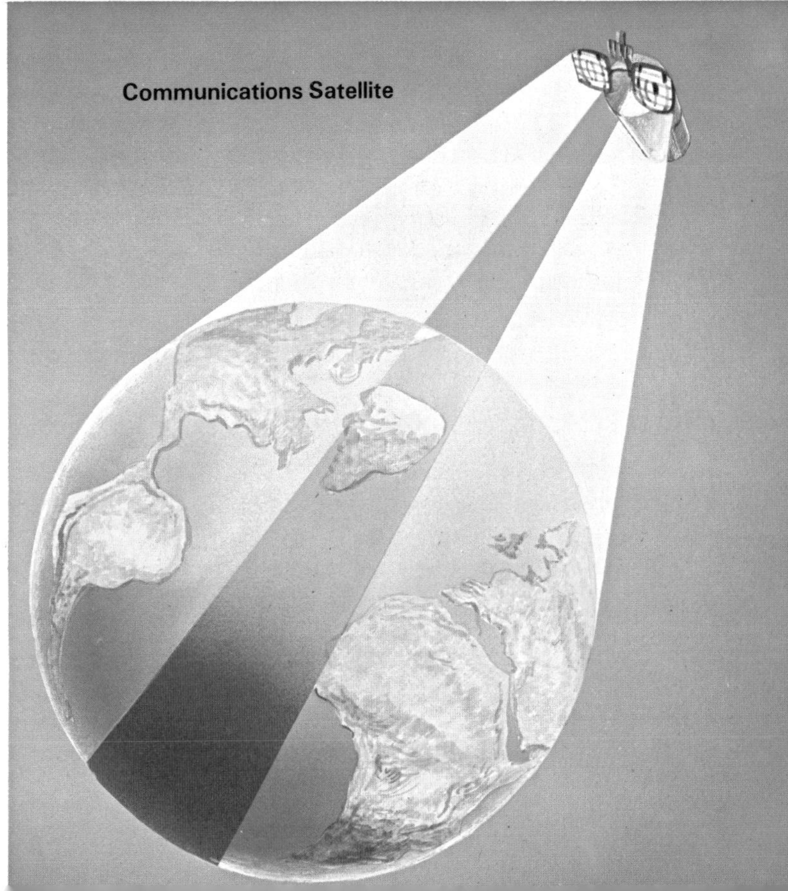

Syncom 3 was the world's first true geostationary communications satellite. It was put into a 24-hour orbit above Earth's equator over the Pacific Ocean in August 1964.
Above: A model of Syncom, showing how the satellite looked in space.
Left: The floodlit gantry surrounding the Delta rocket as Syncom was prepared for its launch.

About one-third of the globe is visible from geostationary altitude, so that a minimum of three satellites is needed to cover the entire globe, each separated by about 120° in longitude. In practice, satellites in the modern international communications network are stationed over the Atlantic, Pacific, and Indian Oceans.

In August 1964, a NASA satellite named Syncom 3 became the world's first true geostationary satellite when it was manoeuvred into position over the Pacific Ocean. (Of its predecessors, Syncom 1 failed to work, and Syncom 2 was put into a 24-hour orbit at an angle of 33° to the equator, so that it appeared to drift from latitude 33° north to 33° south every day.)

INTELSAT

Syncom 3 reached orbit just in time to relay pictures of the Olympic Games from Japan to the US. Its success encouraged the setting up of an international organization called Intelsat (the International Telecommunications Satellite Corporation) to own and operate a global system of communications satellites for commercial use. Intelsat satellites, in geostationary orbit, have revolutionized world communications, from placing an international telephone call to receiving colour TV pictures of world events as they happen.

The Intelsat Corporation includes representatives from most nations of the world, such as the Post Office in the UK, so that no one country owns the network, although various countries have their own ground stations. These ground stations use antennae-like radio telescopes to send up a radio beam to the satellite at one frequency, and receive incoming signals at a different frequency. The radio beams to and from the satellite are called carriers. Signals such as telephone calls or TV programmes are superimposed electronically on the carriers.

Member nations of Intelsat contribute to the cost of building and launching the satellites, and share the profits of their operation. The first satellite owned and operated by Intelsat was Early Bird in 1965, which carried 240 telephone calls or alternatively one TV channel across the Atlantic. Early Bird was the only example of the Intelsat I design of satellite. An improved series known as Intelsat II, consisting of three satellites, was stationed over the Atlantic and Pacific in 1967, thus increasing the range of service.

In 1968 came the first of the Intelsat III series, each of which was capable of handling 1200 telephone circuits or four TV channels at the same time. The Intelsat III series provided a global commercial communications system (military systems with global coverage were already in operation). Intelsat's success brought its own problems, for soon the existing satellites were unable to cope with demand. An entirely new series of communications satellites, Intelsat IV, entered service in 1971 with vastly increased capacity. Each Intelsat IV could carry 5000 telephone circuits or 12 colour TV channels. In addition to antennae which covered the entire visible globe,

All major countries now have ground stations to send and receive signals via Intelsat satellites. They use dish-shaped aerials like radio telescopes, as at these ground stations in Switzerland *(left)* and at Raisting in Germany *(above)*.
Below left: The first Intelsat IV-A communications satellite is readied for launch at Cape Canaveral in September 1975. Satellites like these are stationed over the Atlantic, Pacific and Indian Oceans.

there were special spot-beam antennae focused on specific areas of heavy traffic, such as between the United States and Europe, thus providing increased capacity. Three Intelsat IVs were stationed over the Atlantic, and two each over the Pacific and Indian Oceans.

Even these proved insufficient to cope with demand. An improved series, Intelsat IVA, was introduced in 1975, capable of handling as many as 9000 telephone calls. These satellites are huge, dwarfing a man. Each Intelsat IVA is over 7 metres (23 feet) tall including its antennae, and weighs four-fifths of a tonne. So great is the telephone traffic around the globe, particularly across the north Atlantic, that an entirely new generation of satellites, Intelsat V, is being introduced for the 1980s, using new frequencies and techniques to expand capacity.

Molniya

The Soviet Union has developed its own series of communications satellites called Molniya. Since the most northerly parts of the Soviet Union cannot be served adequately by a geostationary satellite, Russian communications satellites have been put in highly elliptical orbits inclined at 65° to the equator which bring them high over the Soviet Union every 12 hours. A series of such satellites gives national communications without interruption.

Several other nations, particularly those with vast, remote expanses or with heavy telecommunications traffic, have paid NASA to launch satellites for their own national communications. First was Canada with its Telesat series, also known as Anik, the Eskimo world for brother. Indonesia, a nation made up of isolated islands, operates similar satellites, and the United States has followed suit.

In 1978, the European Space Agency launched a prototype communications satellite called OTS (Orbital Test Satellite), forerunner of the European Communications Satellite (ECS) system to be introduced in the 1980s to improve links between European nations and surrounding areas such as oil-rich North Africa and the Middle East.

OTHER ROLES FOR SATELLITES

Communications satellites have proved a great commerical success, but their military role has been equally great. Uninterrupted communications are vital for armed forces, from land units to the crews of missile-carrying submarines. The United States has a network called the Defense Satellite Communications System, and the Soviet Union operates its own counterpart. NATO and the UK have special satellites for keeping their forces in touch around the world.

Staggering new developments in communications are expected in the next decade or so as satellites increase their transmitter power, possibly doing away with the need for large ground stations. Instead, satellites will be able to beam their programmes direct to

Intercosmos 10, a scientific Earth satellite built by the Soviet Union and other Communist countries.

Communication satellites may eventually make possible the wrist radio which will put a person in touch with anyone else in the world.

radio and TV sets. In prospect is a wrist radio which is able to put a person in touch with anyone in the world. Such a system is still somewhat far off, but direct-broadcast satellites are with us at least in experimental form.

In 1975, NASA's Applications Technology Satellite, ATS 6, was stationed over India to broadcast educational programmes to small ground stations in individual communities, bringing information on agriculture, birth control and hygiene. A US military satellite called Tacsat has been used to communicate with individual ships, tanks, and jeeps. No one can foretell the advantages, and possibly disadvantages, of such developments. Direct-broadcast satellites will bring freedom of information and communication to everyone, but they will also allow an endless bombardment of propaganda from the skies. The options are there; how they are used is up to us.

Other specialized purposes for which satellites are used include ship-to-shore communications. In 1976, a series of three satellites called Marisat was launched to link ships on the high seas to their home bases and ensure that distress and rescue calls did not go unheeded. A similar series called Fleetsatcom has been launched by the US Navy to keep in touch with its submarines, ships, and aircraft. The European Space Agency is planning a maritime communications satellite series called Marecs, due for launch in 1980 onwards. Aerosat, an international satellite for aircraft communications, is also under discussion.

Forecasting the Weather

Another way in which satellites have benefited mankind is in weather watching, leading to improved forecasts. Satellites provide meteorologists with pictures of cloud cover over Earth, enabling them to watch the development and movement of weather fronts and storms in far more detail than is possible by the traditional weather reports.

A Tiros weather satellite, one of the series which proved the value of regular weather watching from space.

In particular, in tropical areas where hurricanes can spring up unexpectedly, weather monitoring by satellite can save both property and lives. The savings made in this way are estimated to have paid back the cost of weather satellites several times over.

Tiros 1

The world's first true weather satellite was Tiros 1, launched by the United States in April 1960. The name Tiros stands for Television and Infra-Red Observation Satellite. (Explorer 6, an American satellite launched in August 1959, carried a simple instrument to send back the first pictures of Earth's clouds from orbit, but an unexpected wobble of the satellite blurred the pictures and made them unusable.) Tiros 1 proved its worth as photographs of Earth's cloud cover began to flow in to waiting meteorologists. Tiros 1 and its nine successors orbited Earth every 100 minutes or so at altitudes of about 700 km (435 miles), sending back a total of over half a million photographs, as well as measuring how much heat Earth radiates into space.

A series of larger satellites called Nimbus was launched from 1964 onwards to test out new

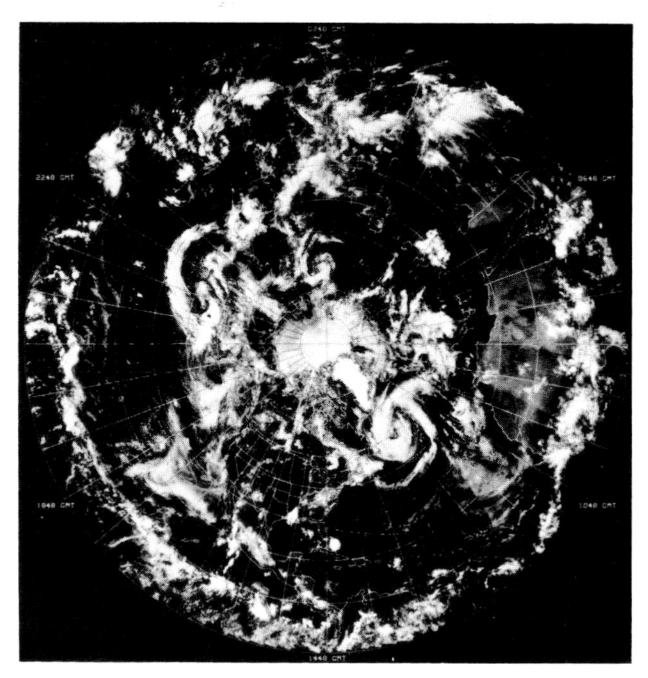

Above: A hurricane photographed by NASA's Synchronous Meteorological Satellite. *Above right:* Clouds over the northern hemisphere seen by NOAA-2.

sensors for studying the atmosphere. These satellites orbited over Earth's poles, so that they passed over each part of the planet every 12 hours, once in daylight and once at night. Cloud pictures can be taken at night by instruments sensitive to infra-red (heat) radiation which is emitted in different amounts by land, sea, and clouds. The Soviet Union has also launched its own network of weather satellites called Meteor, which make similar observations. Data is exchanged between meteorologists of all nations.

As well as returning simple cloud pictures, weather satellites can measure temperatures and humidity at different levels of the atmosphere, and measure the surface temperature of oceans by detecting infra-red radiation and short-wavelength radio waves (microwaves). All this data is vital to weathermen who are trying to make computer calculations of the complex behaviour of the atmosphere.

Snow and ice also stand out prominently on weather satellite pictures. Weather satellites are now being used to track and relay data from balloons in the atmosphere, buoys and ships at sea, and small automatic ground stations in remote areas.

Tiros was superseded by series of satellites known as ESSA and NOAA, after the Environmental Science Services Administration and the National Oceanic and Atmospheric Administration which operated them. These embodied a system known as Automatic Picture Transmission (APT) by which they transmit pictures to any ground stations within range, thus making their results available to all countries.

These satellites were in polar orbits that allowed them to view a given part of Earth every 12 hours or so, but a constant watch on weather over a given part of the globe is also important. This can be achieved by satellites in the much-used geostationary orbit. Weather monitoring from orbit was begun by the ATS (Applications Technology Satellite) series, whose photographs could be put together to make movies of the atmospheric circulation of Earth. Now, the lead of ATS has been followed by several nations, such as the American GOES (Geostationary Operational Environmental Satellite), the European Meteosat and the Japanese Geostationary Meteorological Satellite (GMS).

These and other satellites are contributing to the Global Atmospheric Research Programme (GARP). This is an international collaborative project intended to fill gaps in the existing monitoring of the world's weather and to test the possibility of making forecasts for up to 10 days in advance. In the longer term, we want to know what is happening to Earth's climate–whether the planet is warming up or cooling down, and whether droughts in areas such as the Sahel region of Africa are only localized temporary effects or whether they represent more permanent changes.

We also want to know what effect man is having on the planet, such as by changing the vegetation of large areas and creating new lakes, by releasing carbon dioxide from the burning of fossil fuels, and by releasing other gases from spray cans and high-altitude aircraft into the atmosphere.

Landsats

Wide-ranging surveys of our planet's surface have been undertaken by the Landsats (originally called ERTS, or Earth Resources Technology Satellites), the first of which was launched in 1972. Landsats 2 and 3 followed in 1975 and 1978. The Landsats orbit 900 km (560 miles) above Earth, circling from pole to pole 14 times every day, and taking 18 days to complete one survey of the entire planet. Much larger areas can be covered at a time from orbit than from aeroplanes, therefore orbital surveys are both quicker and cheaper. What's more, geological details show up on space photographs that escape notice from nearer to Earth.

As the Landsats moved along their orbits, they scanned a strip of Earth's surface 185 km (115 miles) wide using two main sensing devices. One system used three TV-type cameras which took pictures simultaneously in three different wavelength bands, from which colour pictures were assembled. The other device, called a Multi-Spectral Scanner (MSS), scans Earth at four wavelengths, from yellow-green to infra-red. Different features are prominent in different regions of the spectrum, such as vegetation which appears bright in the infra-red.

Areas of crop disease or water pollution can be identified by comparing Landsat pictures taken at different wavelengths. Often, images from the MSS are combined to make so-called false-colour composites; in these, the colours used do not correspond to the actual wavelengths at which the pictures were taken, so that vegetation may appear red instead of green. This colour-coding is designed to make certain features more prominent. An experienced geologist can pick out a multitude of features on such false-colour images, ranging from fault lines to sediments, forests and volcanic rocks.

Landsat images have been used

A so-called 'false colour' image of the area of New York City and New Jersey, from NASA's Landsat 1 Earth Survey Satellite.

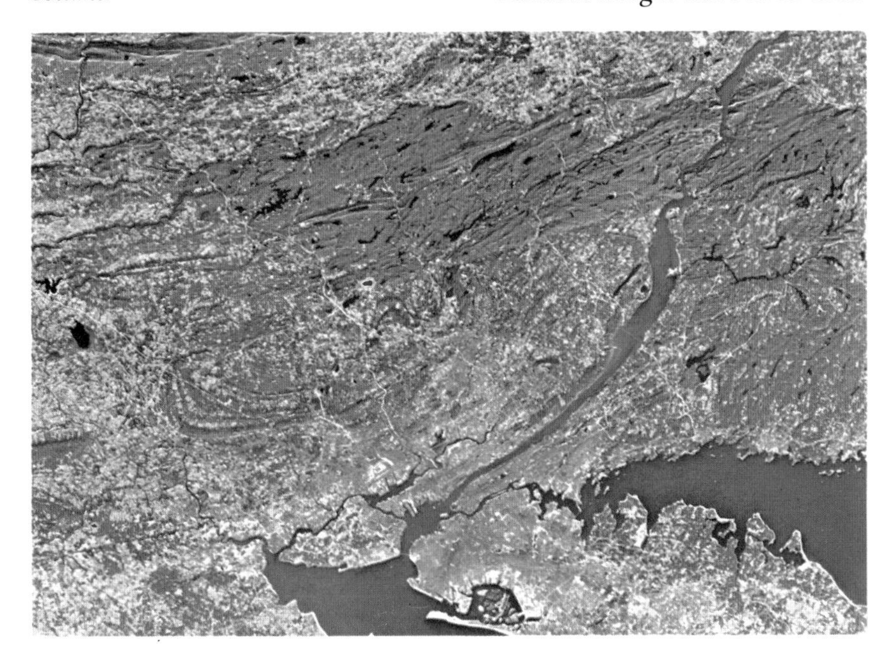

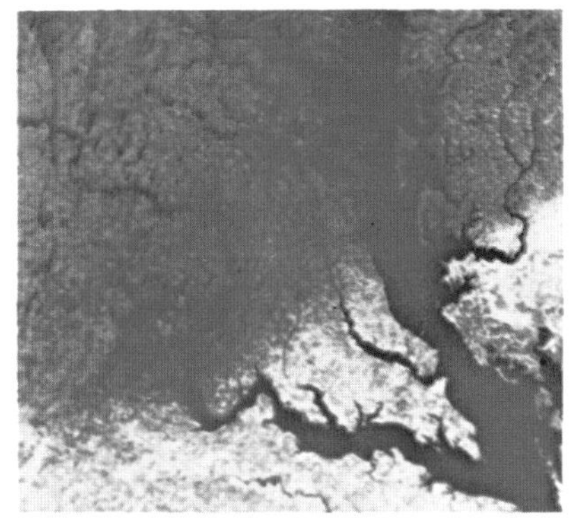

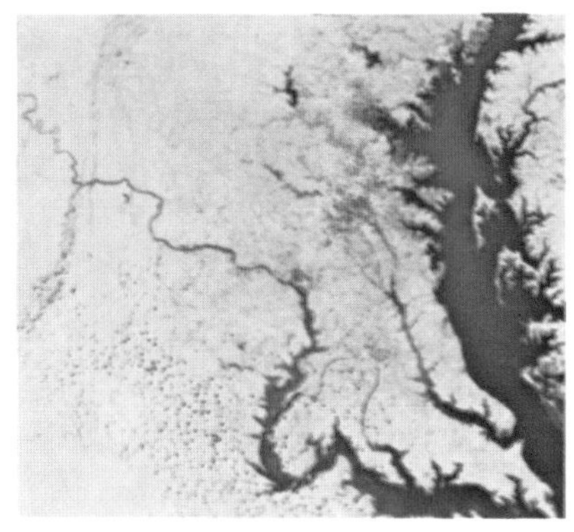

The area of Washington D.C., in *(from top to bottom)* summer, autumn, winter, and spring, as seen by Landsat 1. In these false-colour images, healthy vegetation is shown bright red - note the lack of vegetation in the winter photo and the pink colour in the spring as the vegetation comes back to life.

to identify areas of mineral-bearing rocks, to locate fresh-water lakes in remote areas, and, by some countries, to make more accurate maps than currently available. Such accurate surveys are vital to developing countries. Crop forecasts can be made from Landsat images more quickly, and at least as accurately, as by traditional methods. Ocean currents can be traced by slight differences in colour on the Landsat pictures, as can upwellings which bring nutrients to the surface and thus provide good feeding grounds for fish. Snow and ice cover stand out.

The Landsats, like weather satellites, also act as relay stations for transmitting data from small ground platforms to central receiving stations. Their full impact on monitoring planet Earth is only just beginning to be felt, but it is likely to be substantial. We may begin to wonder how we ever did without such satellites. Experiments have been carried out in space stations such as Skylab and Salyut to improve the sensors used for Earth monitoring, and to widen their range of capabilities. For instance, radar is used to measure waves at sea.

Lageos

A satellite for a rather different kind of Earth survey is called Lageos (Laser Geodynamic Satellite), launched in May 1976. Its purpose is to help measure the drift of continents over Earth's surface. Lageos is a 60-cm (23.6-in) sphere studded with reflectors off which laser beams are bounced from ground stations as the satellite orbits 6000 km (3700 miles) above Earth. This high orbit means that Lageos is essentially unaffected by atmospheric drag, and it is made heavy enough so as not to be affected by the pressure of sunlight.

Since the satellite's orbit is so stable, ground stations can measure their position relative to Lageos to within a few centimetres (or inches). This means slight changes in the station's positions on the face of Earth, as caused by the gradual drifting of continents or even local land movements prior to earthquakes, can be precisely measured. Lageos is expected to remain in orbit for at least eight million years, and carries a plaque showing the location of the continents today and the continents' predicted positions eight million years hence.

Navigation Satellites

One of the less well-known roles of satellites is their use as artificial guiding stars for navigation.

Navigation satellites work by regularly transmitting precise information on their position in orbit, which can be picked up by small receivers on ships or elsewhere. A small computer at the receiving station works out its position relative to that of the satellite. By re-observing the satellite several times, a navigator can fix his position anywhere on Earth to within about 200 metres (660 ft), more accurately than is possible with any other method.

If the receiver is stationary, even better accuracy can be obtained–to within about 10 metres (33 ft). This makes navigation satellites of value to surveyors.

The US Navy was first to show interest in navigation satellites. Its Transit 1, launched in 1960, was the world's first navigation satellite, and has since been followed by a series of similar satellites, mostly classified. The Soviet Union has brought its own navigation satellite system into operation. The Transit series has

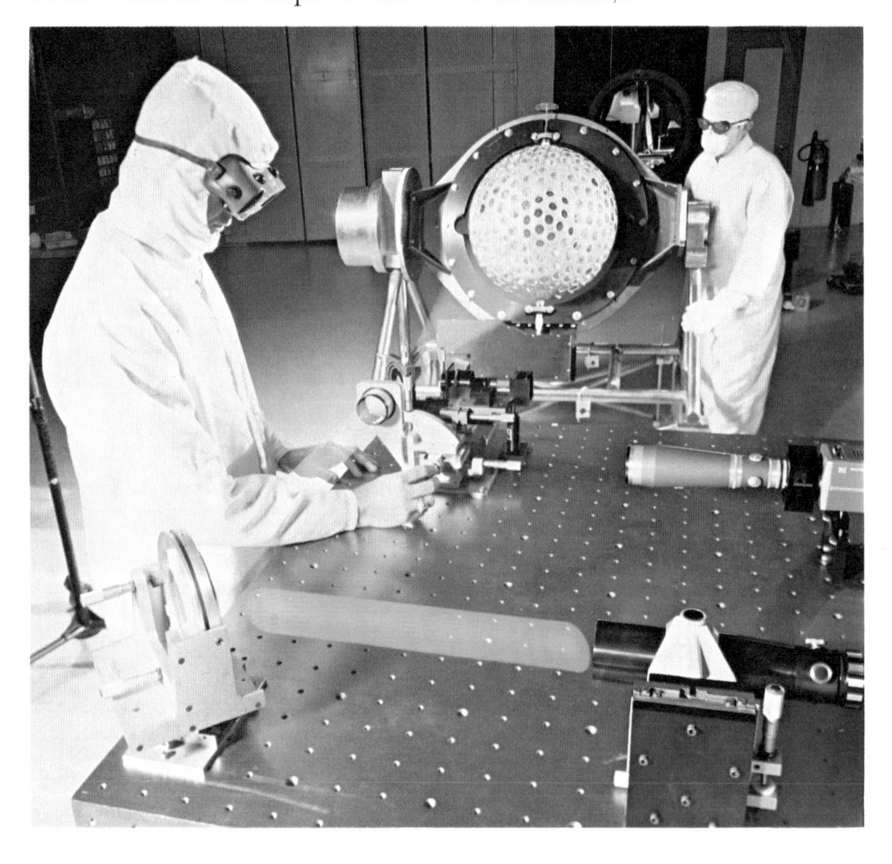

Lageos, a satellite for detecting movements of Earth's crust, under test in the laboratory.

A fourth Earth-surveying Landsat, here shown in an artist's impression, is due for launch in 1981.

been superseded by the so-called Global Positioning System, made up of a network of various satellites in different orbits, available both for navy fleets and civilian ships.

The Transit series was launched into circular orbits about 1000 km (620 miles) up, but satellites in the Global Positioning System are at about 20 000 km (12 500 miles) altitude, giving them a much wider range of visibility.

Improved positional accuracy from navigation satellites helps ships chart more accurate courses and adds to safety in crowded sea lanes, as well as reducing confusion over the limits of territorial waters. Offshore oil rigs use navigation satellites to ensure they are drilling in the correct position.

No one with a suitable receiver need now be lost anywhere on Earth. Satellites in geostationary orbit provide the best visibility of all, and experiments have been undertaken using navigation beacons on NASA's geostationary Applications Technology Satellite (ATS) series.

Scrutiny of the Sun

Pure science has of course benefited enormously from the space age. Satellites provide the best way of examining the upper atmosphere and Earth's surroundings, particularly the magnetic bottle, known as the magnetosphere, which protects Earth from dangerous radiation. Prominent in this field were the American Explorer series and the six Orbiting Geophysical Observatory (OGO) satellites. Astronomers can get a far clearer view of the Universe from orbit than is possible under the dense, turbulent blanket of the atmosphere, and they can observe wavelengths that do not reach the ground, particularly ultraviolet and X-rays.

The Sun has come in for particular attention, notably through the eight Orbiting Solar Observatories (OSO), launched between 1962 and 1975. We rely on our parent star the Sun for the light and heat that governs all life on Earth, so we naturally want to keep as close an eye on it as possible, particularly its minor variations. The OSO satellites

The Sun, our parent star, here seen through telescopes on Earth during a total eclipse, has come under detailed scrutiny from satellites.

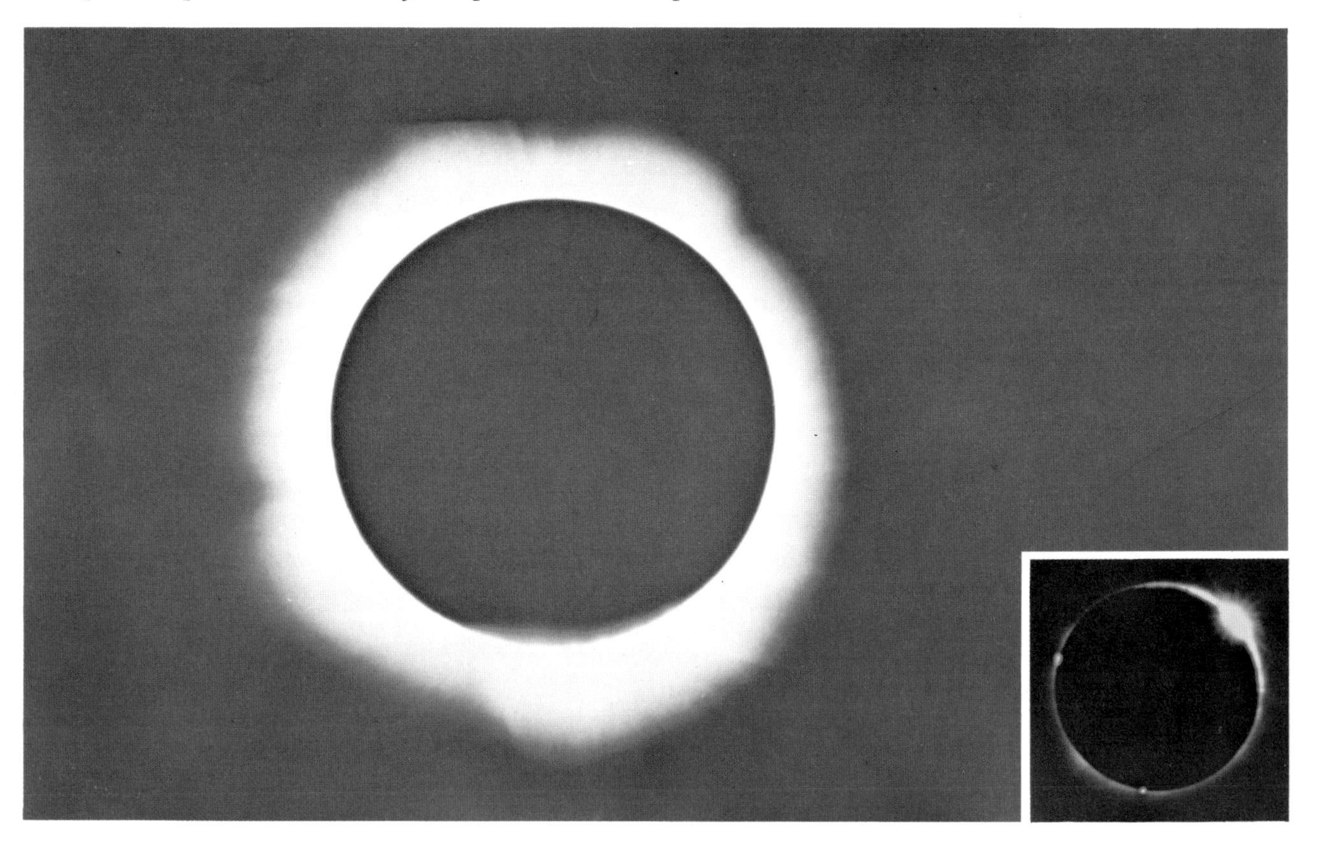

monitored the Sun's activity such as the spots and storms that disturb its surface at intervals.

Solar activity varies with a cycle of about 11 years in length. The OSO series allowed astronomers to follow one complete cycle. Of particular interest were flares, eruptions on the Sun's surface that spew dangerous, short-wavelength radiation into space which can affect Earth's upper atmosphere. From orbit astonomers can also study the faint outer regions of the Sun's atmosphere, notably the corona, which is a thin but intensely hot gas normally invisible from the ground except during total eclipses. Study of the Sun and its surrounding gases was continued in even greater detail by telescopes aboard Skylab.

Telescopes up to 81 cm (32 ins) in diameter have been carried into space for ultraviolet studies of stars aboard satellites of the Orbiting Astronomical Observatory (OAO) series. Hot objects emit most of their energy at short wavelengths such as ultraviolet, and the only way to study these emissions is from above the atmosphere. Among the objects studied by the OAO series were large, hot stars, glowing gas clouds, believed to be the sites of star formation, and other galaxies. The latest satellite to pursue these studies is the International Ultraviolet Explorer (IUE), a co-operative project between astronomers in the United States, the UK, and the rest of Europe; IUE was launched in 1978.

Observatory satellites such as these are producing a revolution in astronomy almost as significant as the invention of the telescope. They show what a limited view of the Universe astronomers have had from the range of wavelengths they have been restricted to on the ground.

A NASA Scout rocket launches Ariel 5, a British astronomy satellite which studied X-rays from space.

X-Ray Astronomy

Of all the discoveries made by observatory satellites, the most exciting are in the field of X-ray astronomy. X-rays are ultra short wavelength radiation emitted by intensely hot gas at temperatures of many millions of degrees. To produce such high temperatures, extreme conditions are required that are unattainable on Earth. Therefore X-ray astronomy allows us to study events that would be impossible to reconstruct in a laboratory. We are glimpsing the behaviour of matter that is literally near the end of its tether.

X-ray astronomy got under way in earnest in December 1970 with the launch of the first Small Astronomy Satellite (SAS-1), also called Uhuru, the Swahili word for Freedom. (SAS-2 studied gamma rays and SAS-3 examined specific X-ray sources in greater detail.) Uhuru made the first complete survey of the X-ray sky, finding 160 sources of X-rays.

Subsequently, other satellites, such as the Astronomical Netherlands Satellite, the British Ariel V, and the first High Energy Astronomy Observatory (HEAO-1), refined and extended this work, finding hundreds more X-ray emitting objects. Most of these are stars or clouds of hot gas in our own Galaxy, but some are galaxies and quasars deep in space. They all have one thing in common: high-energy processes are at work.

What processes are responsible for the production of X-rays in these sources? About two-thirds of the known X-ray sources seem to be twin-star systems in our Galaxy.

What seems to be happening is that gas is flowing from one star to its companion, and heating up as it does so. For the gas to heat up sufficiently to emit X-rays it must be plunging into a super-strong gravitational field, and to produce such a strong field the companion star must be small and dense. The type of star which answers this description is a neutron star, the remains of a formerly large star that has died (see Chapter 1).

But X-ray astronomy allows the detection of even more exotic objects than neutron stars–the famous black holes. A black hole is an object predicted by theory; it has such a strong gravitational field that light cannot escape, therefore it is invisible. But gas falling into this gravitational field around the hole will heat up and emit X-rays, thus giving away the existence of the black hole.

There is at least one X-ray source which seems to contain such an object. The source is called Cygnus X-1, because it was the first X-ray source to be found in the constellation of Cygnus, the swan. Other similar sources containing possible black holes are being found by astronomers.

X-ray satellites have also produced evidence that the energy sources at the centres of super-luminous galaxies and quasars

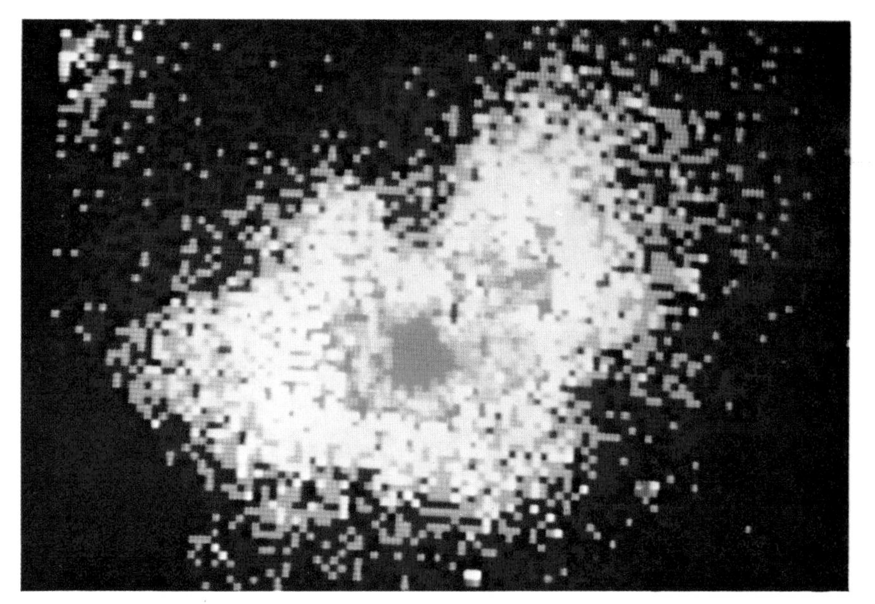

Above: The Space Telescope is due to be launched by the Space Shuttle *(below right)* in the 1980s. *Above right:* The Orion nebula seen in ultra violet light by the International Ultraviolet Explorer. *Below:* HEAO-A, an X-ray astronomy satellite.

deep in space may be powered by massive black holes, formed from the deaths of millions of stars. These giant maws are surrounded by a swirling disk of intensely hot gas that is being swallowed by the celestial drain-plug of the black hole.

Gamma Rays

But even X-rays are not the shortest wavelength radiation being studied by satellites. Gamma rays are shorter still, and are produced by the most energetic of all processes in the Universe.

Some gamma-ray sources have not yet been identified. One particular puzzle is the origin of sudden short bursts of gamma rays, which may indicate the occurrence of violent explosions deep in space. The European COS-B satellite, launched in 1975, has continued the study of the gamma-ray sky. By comparing views of the sky obtained at all wavelengths, from gamma rays and X-rays through the ultraviolet and visible region to radio waves, astronomers will be able to understand much better the processes operating in the Universe.

The Space Telescope

Soon, large optical telescopes will be launched to extend the work of ground-based observatories. Foremost of these is an instrument simply called the Space Telescope, due to be launched by the Space Shuttle in late 1983 or early 1984. It will have a mirror 2.4 m (8 ft) in diameter, allowing it to see details 10 times as fine, and objects 100 times as faint, as the largest telescopes on Earth. The Space Telescope will be pointed from the ground by remote control, and with regular visits from astronauts it is planned to operate for at least 10 years. Its vastly improved performance over existing telescopes promises to revolutionize our knowledge, particularly of the distant parts of space that hold vital clues to both the origin and early evolution of the Universe.

SPIES IN THE SKY

Of course, there are many military uses of space. The majority of space launches are for military purposes – over three-quarters of Soviet launches , slightly less for the United States. Reconnaissance of enemy territory is one of the prime functions.

The United States began experimenting with reconnaissance satellites in its Discoverer series, the first of which was orbited in February 1959. These orbited over Earth's poles, thereby covering the entire globe. One aim of the Discoverer series was to perfect the technique of recovering an instrumented capsule from orbit. This was accomplished with Discoverer 13 in August 1960, making it the first object to be retrieved successfully from space. Discoverer 13's capsule was fished out from the ocean, but subsequent re-entry capsules were snared by aircraft as they floated down under parachutes. The purpose of recovering such capsules was to obtain the high-resolution photographs they contained of ground installations such as foreign air bases and missile sites.

Another type of spy satellite was developed which transmitted its pictures by radio from orbit. Surveys made in this way are naturally not as detailed as those in which the original film is returned, but the results from such satellites can be acquired immediately and they maintain a continuous watch without running out of film. Satellites in this series were initially called Samos (Satellite and Missile Observation System) but, after 1961, the launch of spy satellites became classified and their names were no longer published. Of course, the Soviet Union has also developed classified spy-in-the-sky systems.

Space Wars?

Spying from space might at first seem a provocative act, but in fact spy satellites have been instrumental in controlling the arms race and maintaining world peace. When both sides know what armaments each has, and where they are located, it is impossible for one side to gain a decisive advantage over the other, or to spring a surprise attack. Spy satellites are estimated to have paid back the cost of the entire space programme several times over by preventing unnecessary spending on expensive defence measures.

A new type of American spy satellite entered service in 1971, popularly called Big Bird. These

satellites combine the role of sending back pictures by radio and also, from time to time, ejecting film canisters. Exact details are secret, but the cameras on Big Bird are reputedly able to see objects on the ground as small as a few centimetres (or inches) across. Few activities of any kind around the world escape the attention of the ever-present spies in the sky.

Other military satellites have more specialized purposes, such as monitoring enemy launch sites to give early warnings of missile attack. This was the role of the Midas (Missile Defence Alarm System) series, the first of which was launched in 1960 carrying infra-red detectors to sense the hot gases of a rocket's exhaust. Several advanced early-warning satellites carrying wide-angle telescopes have now been placed in geostationary orbit to monitor and track possible missile launches from the Soviet Union, China, or submarines at sea.

Satellites known as ferrets are used for eavesdropping on enemy radio communications and defence radars, while others use radar and other sensors to track enemy ships and submarines. Yet another type of satellite, called Vela, keeps watch for possible nuclear explosions on or near Earth, thus ensuring that the nuclear test-ban treaty is not violated.

Of all military uses of space, the most disturbing are the so-called hunter-killer satellites being tested by the Soviet Union, which can fly close to another satellite and destroy it. Since 1967 the Soviet Union has demonstrated several times that it can intercept target satellites of its own, put up specially for the purpose. The 'hunter' satellite manoeuvres close to the target and explodes, although sometimes it stays further away as though it were attempting to shoot down the target by powerful laser.

These developments have led to worrying speculations about possible space wars. At present, it is impossible to say whether such a terrifying vision will come to pass, but if war is inevitable then space, well away from earth, is perhaps the best place to hold it.

Space wars – a futuristic battle in the sky.

6 EXPLORING THE PLANETS

Once the first probes had been launched to the Moon, as described in Chapter 3, it was natural to want to take a close-up look at the nearby planets. These neighbour worlds of ours had been studied through telescopes for centuries, but there was much that could never be learned from Earth. Looking back, it is amazing how sketchy, or even wildly off-beam, our ideas were before the first space probes reached the planets.

VENUS–THE HELLISH PLANET

Not surprisingly, Venus was the first target for probes because of its closeness to Earth. It was also a planet we knew very little about, because its enveloping blanket of clouds masked its surface from outsiders' eyes.

Venus is similar in size to Earth, but astronomers did not know whether conditions there would be similar to Earth or vastly more hostile. One charming theory suggested that under its cloud layers Venus was like the Earth in Carboniferous times, with steaming jungles and perhaps even populated with dinosaurs. Another idea said the planet might be covered with soda-water seas, while still another predicted an arid dust-bowl. This latter idea turned out to be nearest the truth, though even that fell short of the full hellishness of the planet.

The world's first successful space probe to another planet was the American Mariner 2, launched towards Venus in August 1962 (Mariner 1, an identical craft, had suffered a launch failure the previous month). Mariner 2 flew past Venus at a distance of 35 000 km (21 748 miles) on December 14, 1962. The probe made no attempt to land and did not carry cameras, but its instruments detected radio waves from the planet which showed that its surface is roastingly hot. Mariner 2 failed to detect a magnetic field or zones of trapped particles like Earth's Van Allen belts.

A planet's magnetic field is believed to be caused by motions in a planet's core as it rotates, but Venus rotates so slowly that sufficient motions are not stirred up to generate a magnetic field. These measurements were extended and refined by Mariner 5 in October 1967.

Sulphuric acid clouds of Venus photographed in March 1979 by the Mariner 10 Venus orbiter probe.

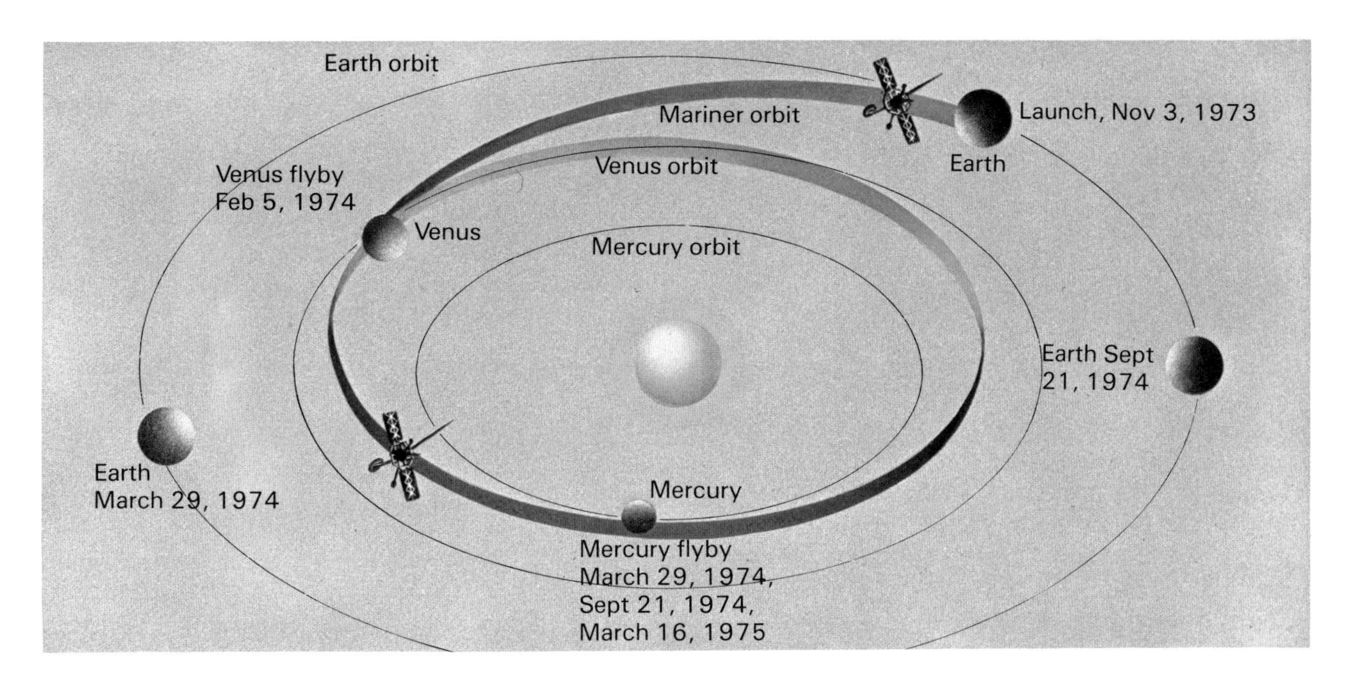

Above: Mariner 10 was the first planetary probe to visit two planets in turn – Venus and Mercury. It used the gravitational pull of Venus to swing it into a different orbit that took it past Mercury three times.
Right: Mariner 2, the first successful space probe to another planet, flew past Venus on December 14 1962.

At the time of Mariner 2, the rotation period of Venus was not accurately known because the surface was invisible. However, radio waves can penetrate the clouds of Venus and radio astronomers on Earth, using giant radio telescopes, began to bounce beams off the planet's surface–a technique known as interplanetary radar.

With radar it became clear during the 1960s that not only did Venus rotate slowly, it took longer to spin on its own axis than it did to orbit the Sun, and the spin was from east to west–the opposite direction to that of Earth and other planets. Space probes have amply verified this surprising result. They have also sampled the atmosphere and clouds of Venus.

Soviet Probe Findings
The Soviet Union has specialized in the space-probe exploration of Venus, although it has been hit by many problems. Contact was lost with its first Venus probe long before it reached its target. In 1966, Venus 2 flew past the planet and Venus 3 actually hit it, but neither returned any data.

Subsequent Soviet Venus probes were more successful. In October 1967 Venus 4 ejected an instrumented capsule into the planet's atmosphere, which floated down under a parachute, measuring the atmospheric composition and pressure. Most of the atmosphere turned out to be made of carbon dioxide, at a much higher pressure than the atmosphere of Earth. But, because of this extreme pressure, the descent capsule of Venus 4 was crushed before it reached the planet's surface, and so readings were cut short.

Two more probes, made of stronger stuff, entered the Venus atmosphere under parachutes in May 1969, but after sending back data during their descent for just over 50 minutes, these, too, yielded to the intense pressure.

A more strongly reinforced probe, Venus 7, made the first successful landing on the surface of Venus on December 15, 1970. Its instruments confirmed that the atmosphere contains over 90 per

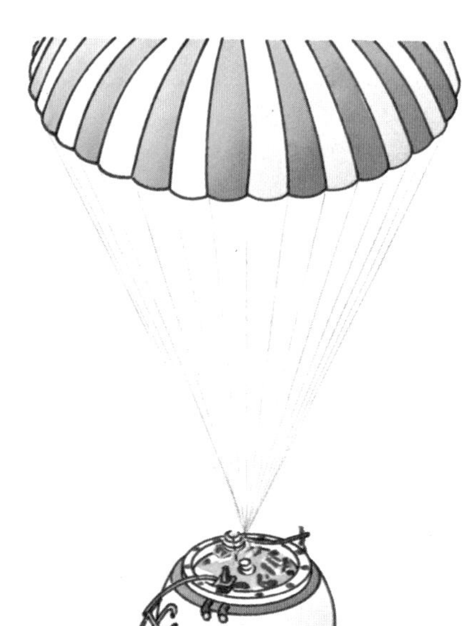

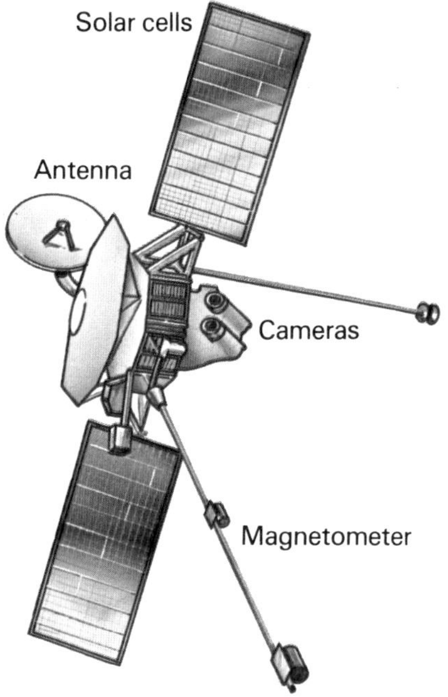

Top: At its brightest, Venus can outshine all but the Sun and Moon.
Top right: A Soviet Venera probe – Venera 4 was the first to land on Venus in 1967. This is the Venera landing capsule which parachutes through the dense atmosphere.
Above: Mariner 10 was the first probe to visit two planets in turn – Venus and Mercury, in 1974.

cent carbon dioxide, at a pressure 90 times that at the surface of Earth.

The Runaway Greenhouse Effect

Temperatures at the surface of Venus are a furnace-like 475°C so that the rocks must glow cherry red with the heat. The high temperature is easily explained by the existence of plentiful carbon dioxide, which traps heat. It is one of the warming agents in our own atmosphere, and there exist fears that releasing too much carbon dioxide by excessive burning of fossil fuels will radically alter Earth's climate. On Venus, the carbon dioxide and a trace of water vapour combine to trap heat from the Sun very efficiently, pushing up temperatures in what is known as the runaway greenhouse effect.

All these probes landed on the night side of the planet, which is the side turned towards us when Venus is at its nearest. In July 1972, Venus 8 parachuted on to the day side of the planet, finding conditions there to be little different from those at night; the reason is that the dense blanket of atmosphere keeps temperatures constant around the planet. Cameras were carried to the surface of Venus for the first time by Venus 9 and 10 in October 1975. They found that sunlight was as strong as on a cloudy day on Earth, quite bright enough to take pictures. Contrary to the general expectation that Venus would be very smooth and eroded under its dense atmosphere, the pictures showed landscapes scattered with jagged rocks. According to a rough analysis by the Venus probes, the rocks are similar in composition to volcanic basalt on Earth.

While the Soviet Union seemed to be concentrating on the surface of Venus, the United States showed more interest in its upper atmosphere and clouds.

First close-up pictures of the clouds came from the American Mariner 10 probe, which flew past Venus in February 1974. Mottled

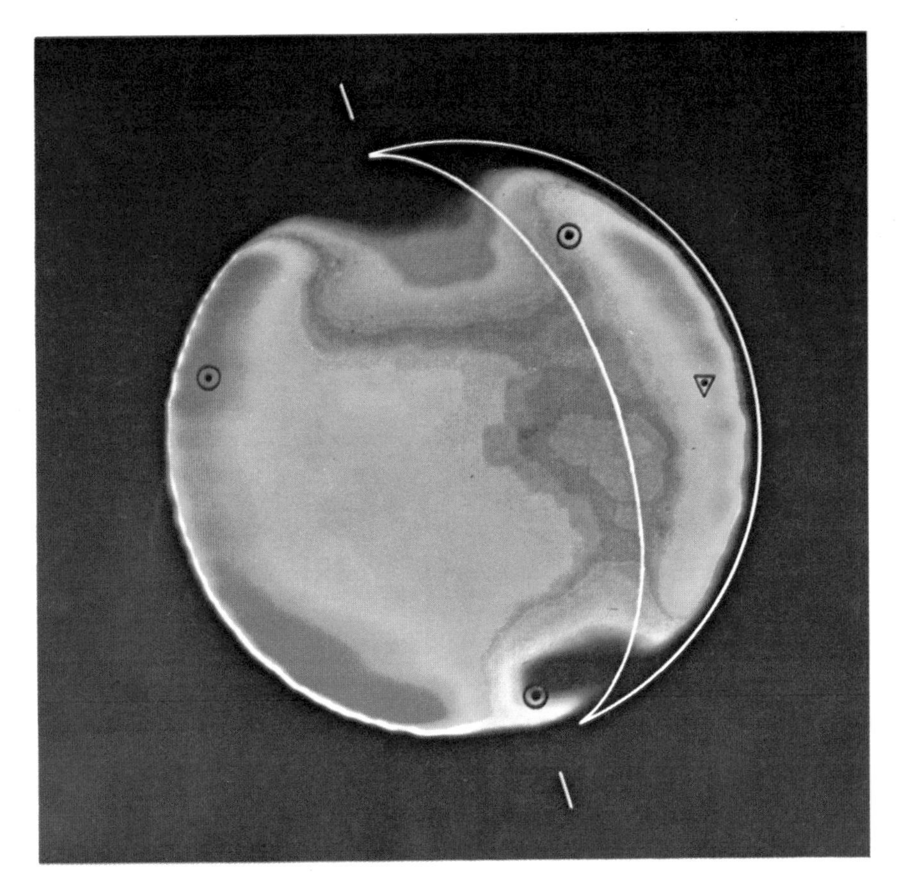

Infra-red image of Venus made from Earth, showing warmer areas (red) and cooler (blue). The planet's sunlit crescent and its poles are indicated in white. The sites of the Pioneer entry probes are marked in black.

regions of cloud are seen over the equator caused by hot gas rising, whereas around the poles the clouds are pulled out into long, smooth spirals. Mariner 10 confirmed observations from Earth that the clouds rotate around the planet every four days, much faster than the body of the planet. Yet the Soviet landers found that wind speeds near the surface of Venus are very slight. Evidently, stratospheric winds of several hundred kilometres (or miles) per hour blow the clouds around the planet. One interesting sidelight of the exploration of Venus is that understanding its atmosphere should tell us more about the behaviour of the atmosphere of Earth.

What are Venus's clouds made of?

It had long been realized that the clouds could not be made of water vapour, as are the clouds of Earth; for one thing, they appear yellowish instead of white. In fact, they are more like smog than true clouds, as confirmed by Mariner 10's photographs which showed them to consist of several layers of haze. The leading recent theory that the clouds consisted of droplets of concentrated sulphuric acid was confirmed in December 1978 by the American Pioneer-Venus 2 as it plunged into the atmosphere. Concentrated sulphuric acid is a highly corrosive substance, thereby adding to the planet's overall nastiness.

During the late 1960s and 1970s radio astronomers were able to make simple maps of Venus, based on radar reflections from its surface. These radar maps revealed apparent craters and lowland basins similar to those on the Moon, numerous mountains including a giant volcano like that on Mars, an apparent lava flow, and a canyon 1500 km (932 miles) long that is comparable to the rift valley of Africa. So Venus is evidently a geologically active planet even today. The radar mapping of Venus was continued in 1979 by the American Pioneer-Venus 1 which orbited the planet; this probe also found that the atmosphere of Venus is rent by incessant bolts of lightning. Hell indeed!

MERCURY–AIRLESS AND CRATERED

While Venus has surprised astronomers, tiny Mercury, closest to the Sun, has turned out to be much as expected: airless and cratered like the Moon. Visual observations of Mercury are difficult because it keeps so close to the Sun, and for a long time astronomers mistakenly assumed that Mercury spins on its axis every 88 days, the same time it takes to orbit the Sun. In the 1960s, radar observations of the planet showed that it actually spins every 59 days, which is two-thirds of its orbital period. Therefore, from one sunrise to the next on Mercury takes 176 days, during which the planet will have orbited the Sun twice and spun three times on its axis.

Dusky markings had been seen on Mercury's surface through large telescopes, and these were presumed to be similar to the *mare* areas on the Moon, but no accurate or detailed maps existed before Mariner 10 encountered the planet in 1974, having flown on from Venus. Mariner 10's photographs of Mercury disclosed a lunar-like surface scarred by craters, some with brilliant ray systems. In fact, at a casual glance it is sometimes difficult to tell the

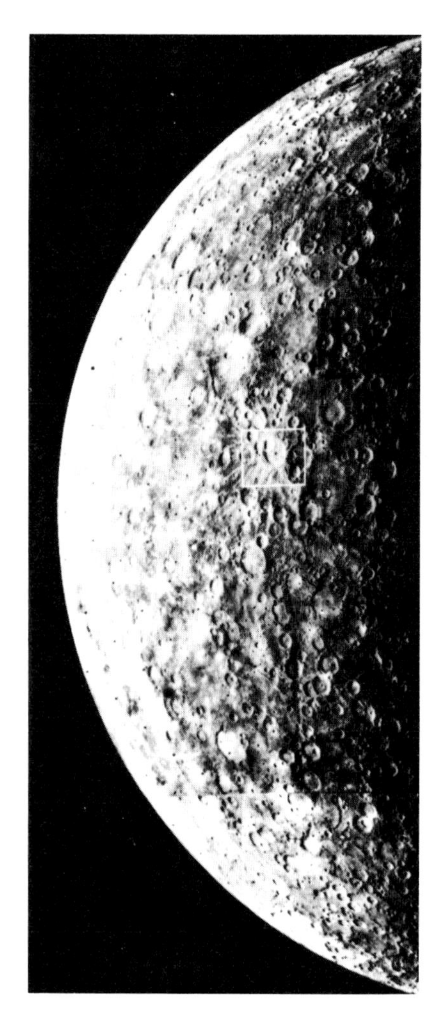

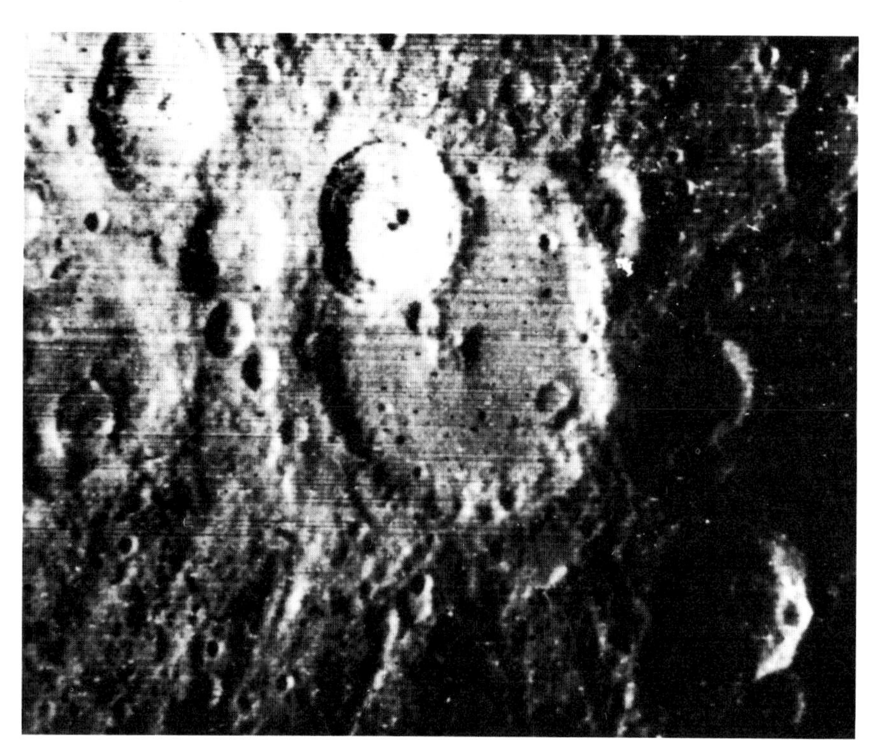

Left: The cratered surface of the planet Mercury, as photographed in 1974 by the US Mariner 10 probe.
Above: This bright crater on the rim of a larger crater is named Kuiper, after an American planetary astronomer. The crater is 41 km (25 miles) in diameter, and can also be seen, above centre, in the crescent view of Mercury on the left.

difference between a photograph of the Moon and a Mariner 10 image of Mercury.

The Caloris Basin

Mercury's most impressive surface feature is a 1400-km (870-mile)-diameter lowland plain called the Caloris Basin, similar to the Moon's giant Mare Imbrium. Lava flows cover its floor, laced with wrinkles and cracks produced as the lava solidified. Mercury seems to have undergone a history similar to that of the Moon, with an intense bombardment by meteorites shortly after its formation followed by a period of gentle volcanic activity as lava oozed into the lowlands.

Its small size, combined with a surprisingly high mass, means that Mercury is an extemely dense body. To account for this high density, Mercury must have an iron core extending four-fifths its diameter.

Why should Mercury have such a large, heavy core? To find the reason, we must go back to the formation of the solar system. Then, the heaviest elements are believed to have been concentrated nearest to the Sun, which is why the innermost planets are solid rock-and-metal bodies while the outer planets are gaseous.

After Mercury formed from dense material near the Sun it must have melted throughout, with the iron going into the core and the lighter rocky material rising to form its crust. Mariner 10 found a slight magnetic field, about one per cent the strength of Earth's magnetic field, evidently produced by the iron core. Pictures from Mariner 10 revealed long cliffs on the surface of Mercury, unlike anything seen on the Moon or Mars. These are thought to have resulted from a slight shrinking of the whole planet, presumably as its iron core cooled and contracted.

LIFE ON MARS . . .?

Of all the planets, Mars has held the greatest popular interest because of the expectation that life might exist there. Through the telescope, Mars seemed quite Earth-like. True, it had no seas, but its day was only slightly longer than ours, it had a thin atmosphere in which clouds formed, and there were even polar caps, albeit thin ones which melted in the summer. Above all, on the red deserts of Mars were dark patches which seemed to change in size and intensity with the seasons. Could these be areas of vegetation?

Some observers, straining to see fine detail on Mars, believed they had mapped a network of narrow

Sunset on Mars photographed by the Viking 1 lander.
Inset: Sunrise over a canyon on Mars known as Noctis Labrynthus. As the ground warms up, overnight frosts evaporate to produce morning mists.

canals, which they presumed had been dug by Martian beings to bring water from the polar caps to irrigate crops (the dark patches) at the equator. Thus arose the vision, more science fiction than science fact, of a dying Martian civilization desperately clinging to survival on a parched planet.

By 1965, when the first space probe reached Mars, these extreme speculations about possible Martian civilizations had been dispelled by improved knowledge of the planet, and the existence of the canals was widely disputed. But it still seemed possible that hardy forms of moss or lichen might exist.

These hopes were dashed in July 1965 by the photographs sent back by the American Mariner 4 probe. These revealed a stark landscape pockmarked by large craters, apparently caused by the impact of meteorites.

Mariner 4 found the atmosphere of Mars to be very thin–less than one per cent the density of our own, giving a surface pressure equivalent to that at a height of 30 km (18.6 miles) above Earth. Air so thin must inevitably be very cold–below freezing point, in fact. It was already known from ground-based observations that there was scarcely any water vapour on Mars; carbon dioxide is the atmosphere's main constituent. Under such hostile conditions, the existence of Martian life seemed almost unthinkable.

Mariners 6 and 7 reinforced this impression of a lunar-like, lifeless Mars when they flew past the planet in 1969. Their photographs

Crescent Mars photographed by Viking 2 showing clouds near a mountain *(left)* and ice patches *(right)*.

showed more craters and large, dust-filled depressions, and their instruments confirmed the dry, frigid, near-airless conditions at the surface. Optimists pointed out that some forms of organism might cling to existence even in such an environment. The only way to find out was to land and investigate.

Phobos and Deimos

In preparation for a landing on Mars, the surface had to be surveyed in greater detail. This was done by Mariner 9, which reached Mars on November 13, 1971, becoming the first probe to orbit another planet. (A sister probe, Mariner 8, was a launch failure, so the entire Mars-mapping job fell to the one craft. It performed exceptionally well.)

Mariner 9 arrived during one of the dust storms that periodically envelop the planet. With nothing to be seen on the surface, the probe turned its attention to the two strange moons of Mars, called Phobos and Deimos. Mere pinpoints of light as seen from Earth, these bodies had intrigued astronomers since their discovery in 1877. Mariner 9's photographs showed them to be irregularly-shaped, cratered lumps of rock, evidently asteroids from the nearby asteroid belt that had strayed too close to Mars in the past and been captured by its gravity. Phobos measures 27 km (16.7 miles) at its longest, Deimos 15 km (9.3 miles). Seeing these two asteroid-type bodies in close-up was an unexpected bonus.

Volcanoes

As the storm on Mars began to clear, Mariner 9's cameras spied three dark spots sticking above the swirling dust. These turned out to be the tops of three volcanoes. Mariner 9 had made the first of its major discoveries that were to radically change ideas about the red planet: Mars *has* been geologically active, and is not the dead planet previous probe results had led us to believe. By chance, the three previous Mariners had photographed the lunar-like half of Mars, missing its most spectacular scenery.

Mariner 9 revealed several other volcanoes on Mars, the largest of which had been seen through telescopes on Earth as a circular white spot named Nix Olympica (the Snows of Olympus). Now known as Olympus Mons (Mount Olympus), this object turns out to be the largest volcano in the solar system, larger even than the volcanic Hawaiian islands on Earth. It is 600 km (373 miles) wide, 25 km (15.5 miles) high, and is topped by a complex summit crater. Just as spectacular, and unexpected, was the so-called Martian Grand Canyon, actually a rift valley up to 4000 km (2485 miles) long and 120 km (74.5 miles) wide, evidently a fracture

Models of Mars made from space probe photographs, showing part of the great Martian rift valley *(above right)*, the massive Martian volcano Nix Olympica *(below)*, and a close-up of Nix Olympica's complex summit crater *(right)*.

in the planet's crust that has subsequently been eroded by wind-blown dust.

The existence of volcanoes on Mars meant that the planet's atmosphere had probably been denser in the past. Volcanic gases are believed to be the source of the present-day atmospheres of Earth, Venus, and Mars, although in the case of Mars some of the gas has subsequently escaped. A denser atmosphere means warmer climate and possibly water (a large proportion of volcanic gas is water vapour). As if to prove this point, Mariner 9 discovered long, winding valleys on Mars, which look like dried-up river beds. When the volcanoes erupted in the past, the climate of Mars may well have been suitable for life to begin.

Mariner 9's pictures laid two controversies to rest. Firstly, it became clear that the seasonal changes in the dark markings of Mars are due to the effects of wind-blown dust, which uncovers or obscures areas of dark rock; the changes are definitely not due to growing vegetation.

Secondly, of the notorious canals there was no sign. Some elongated dark markings and other smudgy features on the surface could be equated with certain of the canals drawn by telescopic observers, but the majority of fine lines painstakingly mapped as 'canals' have no related features. The canals existed only in the imaginations of certain observers. who must have been the victims of optical illusions caused by disjointed surface markings seen under less than perfect conditions.

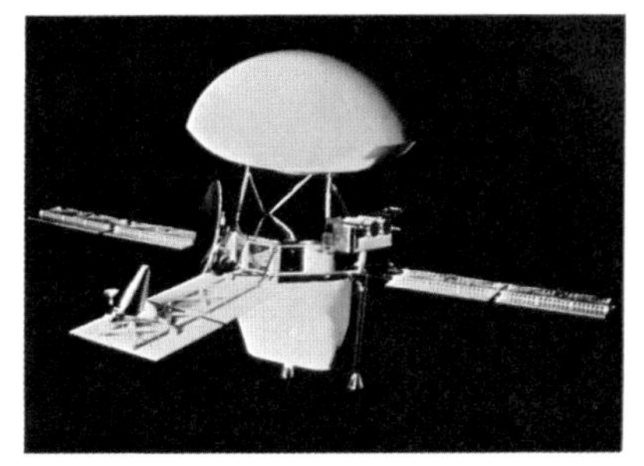

Above: Viking 1 being assembled in a clean room at Cape Canaveral prior to its launch in August 1975. *Right:* Viking as it appeared in space. The white dome contains the lander. The orbiter is the section with the windmill-like solar panels.

Viking Invasion!

Space-probe exploration of Mars reached its climax in 1975 with the launch of two Viking spacecraft. Each of the identical Vikings came in two halves: an orbiter to survey the planet from above, and a lander to perform experiments on the ground. Landing on Mars was expected to be difficult. A warning had been sounded by a small fleet of Soviet probes which had failed in previous landing attempts. It was theorized that the lander capsules had been blown over by unexpectedly strong winds or had toppled on rocks or drifts of soft dust.

On the basis of the Mariner 9 pictures, smooth, flat areas were selected as Viking landing sites. But, once the Vikings reached Mars and began to survey the selected landing sites from orbit, it became clear that these areas were rougher than anticipated. Last-minute changes of landing sites had to be made before the first Viking lander began its descent under parachutes, using rockets to brake its final touchdown.

It came to rest on July 20, 1976, in a lowland area called Chryse (rhymes with 'icy'), over which liquid water was believed to have flowed in the past when Mars was wetter.

Two cameras on the Viking lander began taking colour photographs of the surroundings, revealing a red, rocky landscape reminiscent of some deserts on Earth. As expected, the red colour of Mars turned out to be caused by

Sunrise on Mars photographed by Viking lander 2 in June 1978 — nearly two years after its touchdown. The Sun lies just above the horizon.

large amounts of iron oxide (rust!) in the rocks. All that was missing was any sign of life, either plant or animal. Tiny patches of vegetation, or even animals burrowing or hopping on the surface, would have been caught by the Viking lander's cameras. Despite careful scrutiny over many months, no changes due to any form of life were seen.

Martian Surprises

One immediate surprise was the colour of the Martian sky: it is pink, caused by fine particles of sand suspended in the atmosphere. Also unexpected was the large number of rocks and boulders scattered around, evidently thrown out from meteorite impacts. Viking had been lucky not to crash-land on one of them.

Viking 1's lander came down in temperature latitudes, about 22° north of the Martian equator. Viking lander 2 came down further north, in a region called Utopia, at latitude 47°, on the edge of the area covered by winter frosts. Its cameras showed a red, rock-strewn landscape similar to that at the Viking 1 lander site.

Instruments aboard each Viking recorded mild breezes blowing at both landing sites. The maximum air temperature at Chryse was −29°C in mid-afternoon, falling to −85°C before dawn–and this in midsummer! At neither Viking site does the temperature exceed freezing point at any time of the year. In winter, carbon dioxide frost collected on the ground around the Viking 2 lander in Utopia.

Analysis by the landers of the atmosphere of Mars revealed the existence of two or three per cent of nitrogen, the first time this gas had been detected on the planet. Nitrogen is one of the three vital ingredients for life, the two others

Red sands of Mars, photographed by the Viking 1 lander. At bottom right are trenches dug by the Viking's soil sampling arm in search of life.

Rock-strewn surface of Mars at the Viking 1 landing site. Largest rocks are about 30 cm (1 foot) across.

being carbon and water, already known to exist on Mars. The polar caps of the planet are made of a mixture of frozen water and frozen carbon dioxide. There is believed to be a layer of perma-frost under the whole of the planet's surface. Mars therefore does not lack water entirely; the water simply cannot become liquid under present-day conditions of freezing temperatures and low air pressure.

Testing for life

Inevitably, most interest centred on the Viking experiments to detect life. Even though nothing living was visible to the Viking cameras, tiny bugs like bacteria might have existed in the soil. Each Viking carried a miniature biological laboratory in which soil samples collected by the lander's scoop were incubated in three different ways to encourage any Martian micro-organisms to grow.

In one test, the soil was fed with a rich nutrient solution, colloquially termed 'chicken soup', and the gases given off were analyzed. Carbon dioxide and oxygen were both released from the soil in this experiment, but the results were not thought to be due to life. Instead, the release of the gases has been attributed to chemical reactions between the nutrient and the soil.

In a second experiment, a liquid nutrient containing radioactive carbon was injected into the soil. Carbon dioxide containing the radioactively-labelled carbon was given off by the soil in this experiment, but again the result was attributed to chemical reactions between the soil and the nutrient, not life. The Martian soil is found to be highly oxidized (containing a lot of oxygen), and this oxygen was evidently combining chemically with the carbon of the nutrient to produce carbon dioxide.

Thirdly, soil samples were incubated under an atmosphere containing radioactively-labelled carbon dioxide to see if anything in the soil might take in carbon, as plants would be expected to do by photosynthesis. Analysis of the soil showed that radioactive carbon was indeed being taken up from the atmosphere, but not in such a way as to convince experimenters that life was present. Probably, the results of all life-seeking experiments at both Viking sites were due to chemical reactions with the soil, not the growth of Martian micro-organisms.

A related experiment aboard the Vikings, which would have given clues to the existence of life on Mars, also produced negative results. This experiment analyzed the soil for organic molecules–those which would make up the bodies, and the food, of any Martian micro-organisms. No such molecules were found –only carbon dioxide and water.

Therefore it seems that, despite the hopes of optimists, there is no life on Mars in the two locations of the Viking landers. Life may exist in certain favoured oases on Mars, possibly nearer the polar caps, but the chance is slim.

More American space probes may go to Mars to extend the search for life. One proposal is for a Mars rover which could roam the surface on caterpillar tracks in search of the most favourable niches for life. The Soviet Union is also believed to have plans for a similar craft, based on their Lunokhod lunar rover.

Most exciting of all is the idea of bringing back a sample of Mars soil by automatic probe, in the same way that Soviet lunar landers have brought back soil from the Moon. But a Mars sample return is unlikely to take place before 1990. On this schedule, men will not reach Mars until well into the next century.

GIANT JUPITER

Next planet in line from the Sun is giant Jupiter. Its exploration is only just beginning because, despite its interest, it is much further away than any of the inner planets which have been probed so far.

Jupiter is in many ways a midway stage between a true planet and a small star, for it is made largely of hydrogen and helium gas, similar to the composition of the Sun. Jupiter has a very intense magnetic field which stretches far out into space, trapping charged atomic particles to form intense zones of radiation like Earth's Van Allen belts.

As seen from Earth, Jupiter's most striking feature is its multicoloured clouds, which are drawn out into horizontal belts by the planet's swift rotation every nine hours 55 minutes. The colours of these clouds range from red to blue, but they usually appear yellow or brownish. The cause of the colours must be various chemicals mixed in with the main gases of Jupiter. It has even been suggested that chemical processes like those that led to life on Earth in the distant past are still taking place on Jupiter today (see Chapter 9).

No one has seen below the clouds of Jupiter so we can only speculate what the planet is like inside. The clouds around Jupiter are believed to be about 1000 km (620 miles) deep. As we go down through this layer the clouds become denser and warmer; at the lowest levels, the clouds may be composed of water droplets, but the higher clouds are frozen ammonia crystals. Below the cloud layer, Jupiter is composed almost entirely of liquid hydrogen, compressed by the giant planet's strong gravity. At Jupiter's centre may be a metallic and rocky core about the size of Earth. Convection currents in the liquid hydrogen around the core are believed to be responsible for setting up the planet's strong magnetic field.

The Great Red Spot

Jupiter's cloud formations are constantly changing, except for one feature: the great red spot. This is a ruddy-coloured oval large enough to swallow several Earths, and has been seen through telescopes for at least 300 years. Its cause remained uncertain until quite recently. At one time it was thought to be a raft of semi-solid material floating in the clouds, but now it is known to be a swirling eddy of clouds. Its colour remains a puzzle; one theory attributes the hue to red phosphorus being wafted up by warm gas rising from deep within the clouds. The atmospheric circulation of Jupiter is driven by heat from inside the planet. Probably Jupiter is still cooling off from its formation.

Below: Jupiter and its moon Io from Voyager 2.
Below right: Rings of Jupiter show orange on this photograph by Voyager 2. The picture was taken through colour filters, giving Jupiter a blue and red rim.

Above: Great red spot in Jupiter's clouds photographed by Voyager 1. Next to it is a white oval cloud.
Above right: Voyager 1 close-up, with blue colouration enhanced, shows complex structure in the red spot.

Meteorologists, in particular, were keen to get better views of the clouds of Jupiter. Two probes, Pioneer 10 and 11, were launched towards the planet, flying past it in December 1973 and December 1974 respectively. As the Pioneers penetrated Jupiter's magnetic field they were subjected to doses of radiation hundreds of times stronger than that needed to kill a man. They provided better pictures of the turbulent clouds than available from Earth, confirming that the bright zones are caused by ascending gases, and the dark bands are deeper layers where gas is descending. In addition to Jupiter's great red spot, the Pioneers photographed some smaller, shorter-lived red spots, which appear from time to time in the clouds.

Pioneer 10 passed the planet's equator but its sister craft Pioneer 11, on a different trajectory, gave astronomers a view of the planet unobtainable from Earth when it flew over Jupiter's poles. Here, the familiar pattern of bands and streaks is less pronounced, because rotational forces are lower. Instead, individual convection cells of rising gas were visible as mottled patches.

Voyager's Surprise Findings

Jupiter was surveyed in even more detail in 1979 by the Voyager 1 and 2 probes. Their photographs were astounding, revealing fantastic looped and scalloped edges to the cloud bands, as well as spots and swirls like high and low pressure systems on Earth. Never before had the full complexity of Jupiter's clouds been so clearly displayed. Picture sequences of the great red spot confirmed it to be a rotating eddy of cloud, twisted around by the planet's rotation. One surprise was Voyager 1's discovery of a faint ring of particles orbiting Jupiter, too faint to be seen from Earth.

Most spectacular of all, Voyager 1 photographed active volcanoes on Jupiter's moon Io,

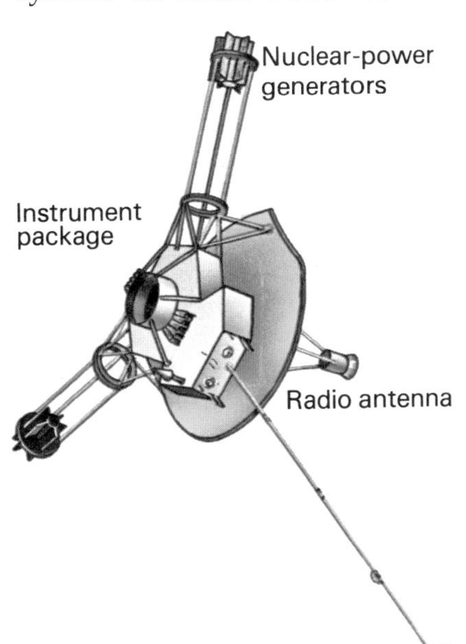

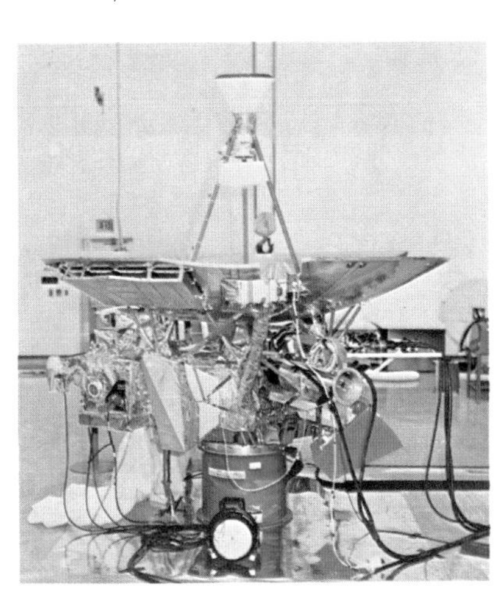

Below left: Pioneer 10 flew by Jupiter in December 1973 and was followed a year later by the near identical Pioneer 11.
Below: A Pioneer spacecraft during assembly.

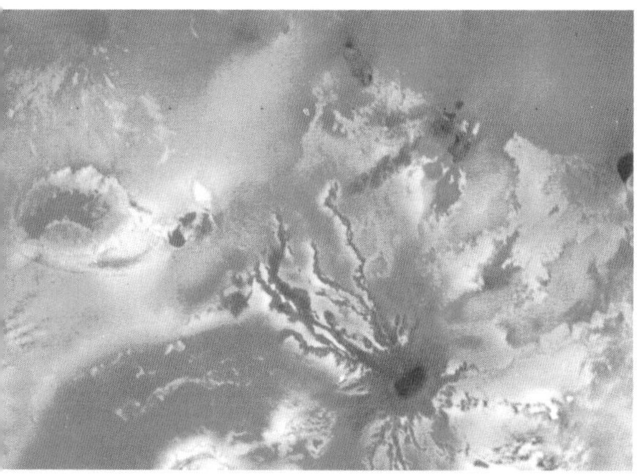

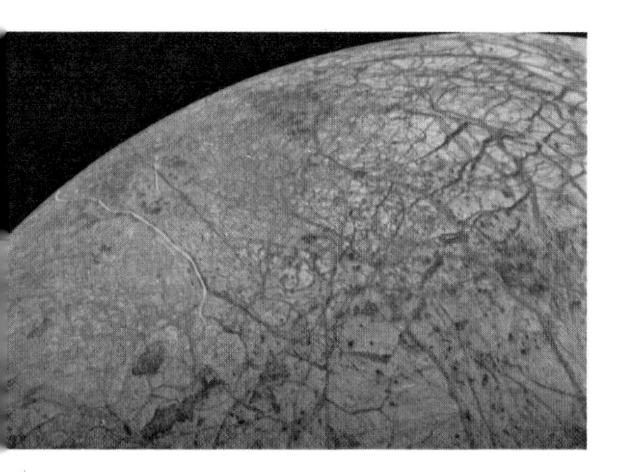

Top: A volcano erupts at the limb of Jupiter's orange-coloured moon Io.
Middle: Red and orange coloured flows of sulphur and salt surround volcanoes on the surface of Io.
Above: Fractures in the icy crust of Jupiter's moon Europa.

the surface of which is covered with yellow and orange patches, probably deposits of salt and sulphur. Io is surrounded by a glowing cloud of sodium. On other moons of Jupiter, notably Callisto and Ganymede, apparent meteorite craters and ice fields were seen. Future exploration of Jupiter will include an orbiter and probes to plunge into the atmosphere, sampling its composition, during the 1980s.

Saturn and its rings photographed by Pioneer 11 in August 1979. Saturn's moon Rhea is seen at bottom right.

SATURN CLOSE-UP

After passing Jupiter, Pioneer 11 went on to take the first close-up look at Saturn, arriving there in September 1979. Saturn is in many ways similar to Jupiter, particularly in chemical composition, although its clouds are neither so active nor so colourful. Saturn has the famous rings which are composed of countless particles of rock. Astronomers want to know more about the particles that make up these rings, and to check the existence of small moons orbiting at their edge.

The two Voyager probes which passed Jupiter in 1979 were also scheduled to fly on to Saturn, arriving there in 1980 and 1981. One of the prime targets for the Voyagers was a close-up look at Titan, Saturn's largest moon, which is larger than the planet Mercury and has a dense atmosphere. Voyager 2 is also expected to fly on to look at Uranus, arriving there in 1986, and perhaps even to Neptune, which it would not reach until 1989. Very little is known about these remote planets.

Plans for future exploration of the solar system by space probes have been curtailed by cuts in NASA spending. Other nations are expected to assume an increased role in planetary exploration in the 1980s, such as European countries and Japan.

One plan is for a probe to rendezvous with a comet, such as the famous Halley's comet which is due to return in 1986. Comets are bags of rock and gas left over from the formation of the solar system and, by studying them, we may hope to get a better understanding of the nature of the original material which gave birth to the planets, including our own Earth. No one has seen a comet from closer than a few million kilometres (or miles), so we stand to learn a great deal about these ghostly wanderers.

7 A NEW ERA IN SPACE

Manned spaceflight is entering a new era. Following the pioneer flights into orbit and the first explorations of the Moon, attention has turned to using space for practical benefits on Earth–not just in surveying the sky and land, but also in the more far-reaching prospect of putting industry into orbit. Experiments have been conducted in the American Skylab and the Soviet Salyut space stations, utilizing the unique conditions of weightlessness and vacuum to manufacture products such as new alloys or valuable types of glass that are unobtainable on Earth. Men are also learning to live and work for long periods in space.

We are seeing the start of a new industrial revolution–a revolution which promises to make all the expenditure on astronautics look like one of the best investments in history. In addition, it is the first step in mankind's colonization of space, which may lead to the eventual spread of human life throughout the Galaxy.

A futuristic spaceship. Will this be the normal way to travel in the future?

THE SPACE SHUTTLE

Key to the new era is the re-usable space transportation system popularly termed the Space Shuttle. The Shuttle is a winged craft the size of a large jetliner. It is launched like a rocket but can fly back to land on a runway like an aircraft. It carries its payload in a cargo bay and, after its return, this bay can be filled with a new payload and the Shuttle readied for another launch.

The Shuttle's re-usability means that it can cut the cost of launches by up to 90 per cent over conventional throw-away rockets. And it's this reduction of launch costs, along with the Shuttle's ability to bring freight back from orbit in its cargo bay, which makes space manufacturing an attractive prospect. Larger and more economical shuttles may come into operation once space industries become established.

The Shuttle's activities in launching men and satellites will speak for themselves when it comes into full-time operation during the 1980s. For the moment, let's concentrate on the vehicle itself, and also look at mankind's first attempts at long-term living in space stations.

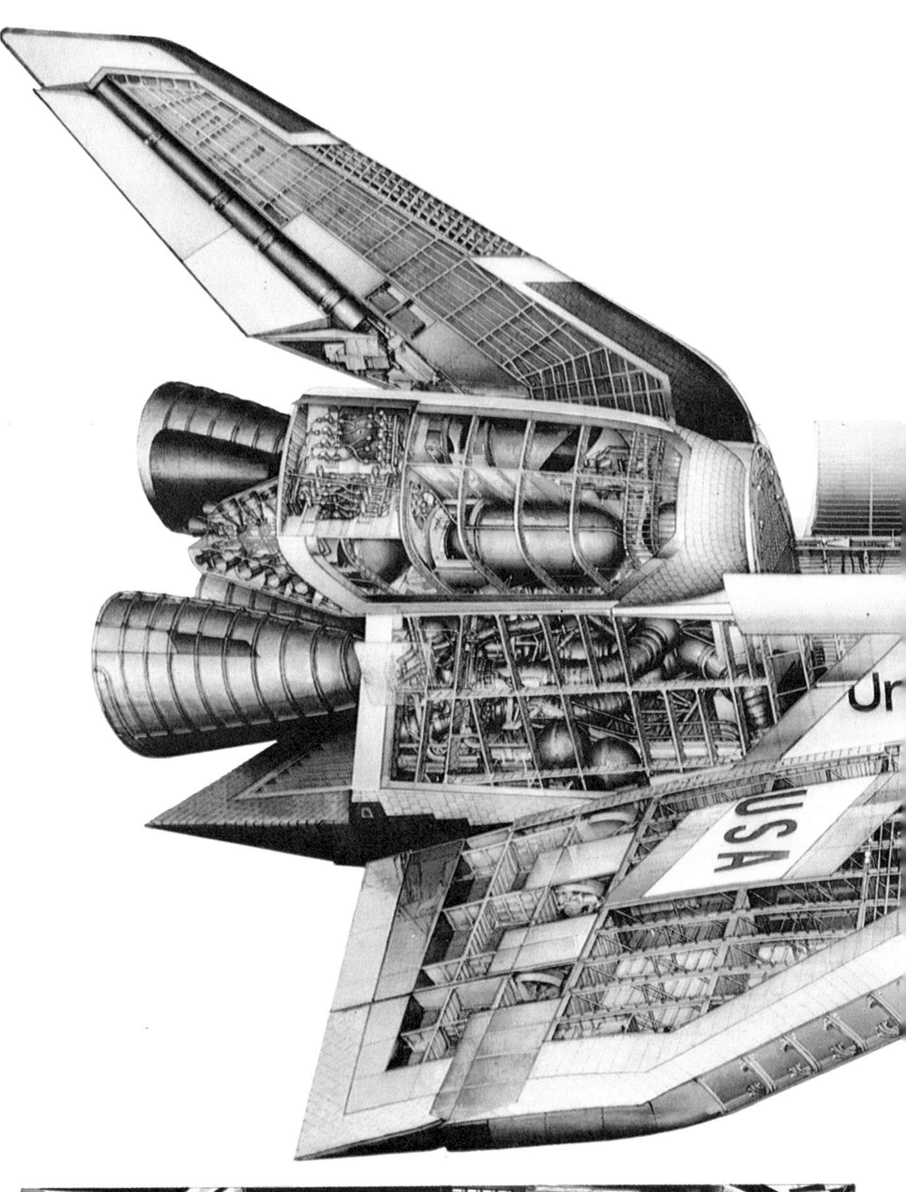

The X-15

The concept of a space shuttle is not as new as it might seem. Eugen Sänger (1905–1964), an Austrian spaceflight pioneer, originated the idea of a winged, re-usable rocket plane in the 1930s, which he foresaw would ferry up men and materials to orbiting space stations. Following the Second World War the United States carried out experiments with high-altitude rocket planes. These resulted in the famous X-15 which, during the 1960s, flew at altitudes up to 108 km (67 miles) and speeds over 7000 kph (4350 mph).

The X-15 took man to the edge of space and, had it been launched atop a carrier rocket, might have put a man into orbit. But by then the task had already been done by simpler, non-re-usable craft. NASA in the 1960s carried out tests with wingless but aerodynamically-shaped high-speed gliders called lifting bodies, which were dropped from aircraft and piloted to a runway to simulate return from orbit. Experience with these craft proved valuable in developing the modern Shuttle.

NASA began its plans for the Shuttle in 1968. As originally conceived, it was to consist of two stages, both winged and both piloted. The orbiter rode piggy-back on a massive first-stage booster. The winged booster was to be flown back by its crew for re-use, while the orbiter went on into space. Such a spectacular two-stage manned vehicle would

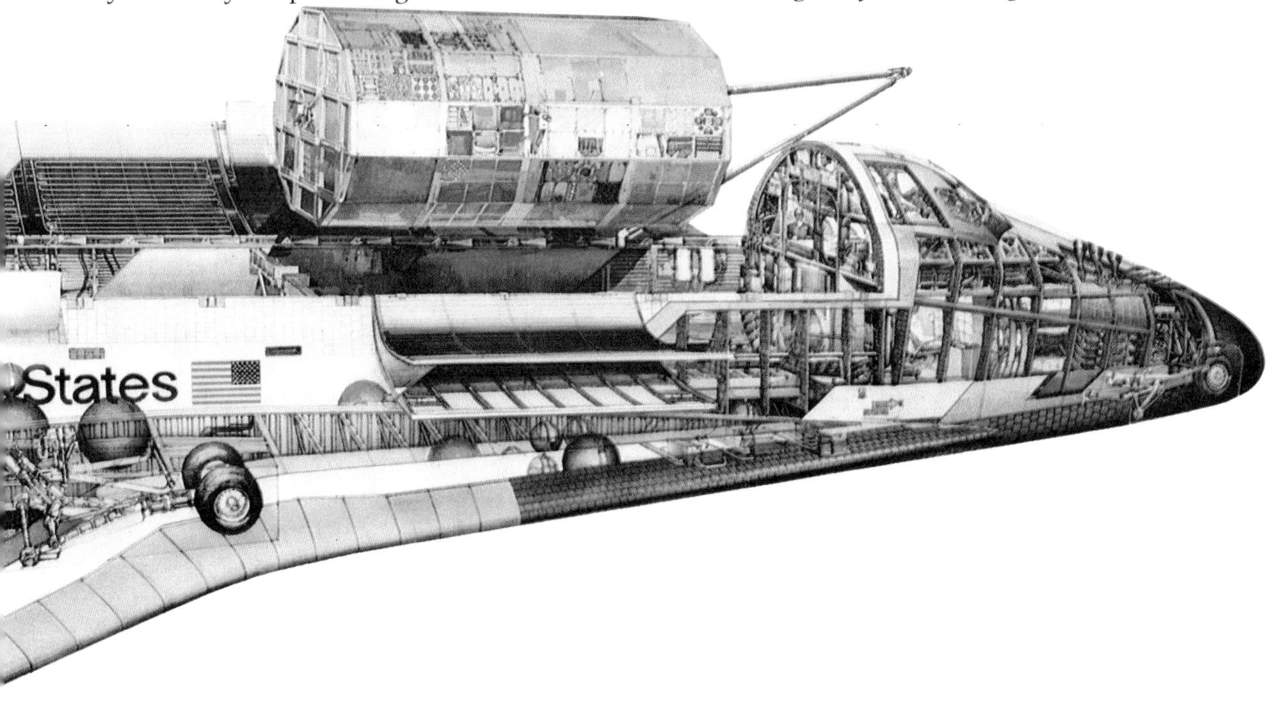

Far left: The Orbiter's flight deck showing the main instruments.
Left: Interior of the Shuttle flight deck showing the four-person crew as they will be seated for launch and landing.
Left below: Artist's impression of astronauts at work on the flight deck of the Space Shuttle while in orbit.
Below: The X-15 was an experimental rocket plane that took men to the edge of Earth's atmosphere in the 1960s.

The Space Shuttle, America's re-usable space plane of the 1980s, shown in cutaway. At rear are the three main engines used during launch; these are fed by fuel from an external tank which is discarded once in orbit. The Shuttle has a wheeled undercarriage for landing on a runway on return from space. At front is the flight deck with, underneath it, living quarters for the Shuttle astronauts. A remote controlled arm is shown lifting a satellite out of the payload bay. The Shuttle manoeuvres in space by use of small rockets in the nose and tail.

have proved costly and difficult to develop, so the Shuttle's final design was a compromise, relying on conventional rockets to assist its take-off. Nevertheless, in the true spirit of economy, these boosters are intended to be recovered by parachute and re-used.

The Orbiter is a delta-wing craft 37.2 m (122.5 ft) long and with a wingspan of 23.8 m (78 ft). In its forward compartment ride a crew of three and up to four passengers. This pressurized section contains a galley, toilet, sleeping quarters, and general accommodation. Most of the Shuttle Orbiter's length is taken up by the cargo bay, 18.3 m (60 ft) long and 4.6 m (15 ft) wide, which can carry a payload of 29 tonnes into orbit, and bring back to Earth up to 14.5 tonnes. Doors on the cargo bay open up once the Shuttle is in space. One of the Shuttle's roles will be to retrieve unwanted or malfunctioning satellites for disposal or repair.

Above: Flight testing the Shuttle Orbiter in the atmosphere on the back of a jumbo jet.
Left: The Shuttle on the launch pad at Cape Canaveral.

How the Shuttle Works

This is how the Shuttle will operate: at launch, the Orbiter is mated to a large external tank which supplies fuel for the Orbiter's three main engines; the Orbiter itself carries no fuel except for its small manoeuvring engines. To this tank are attached two solid-fuel boosters. All engines run at lift-off and the Shuttle slowly ascends; its launch power is second only to that of the Saturn V. At an altitude of 43 km (26.7 miles), the two solid-fuel boosters burn out and drop away, descending under parachutes to be retrieved from the ocean, refilled with propellant, and used again. Under continuing power from its own main engines, the Orbiter ascends into space, the external fuel tank being jettisoned once in orbit. Of the whole Shuttle

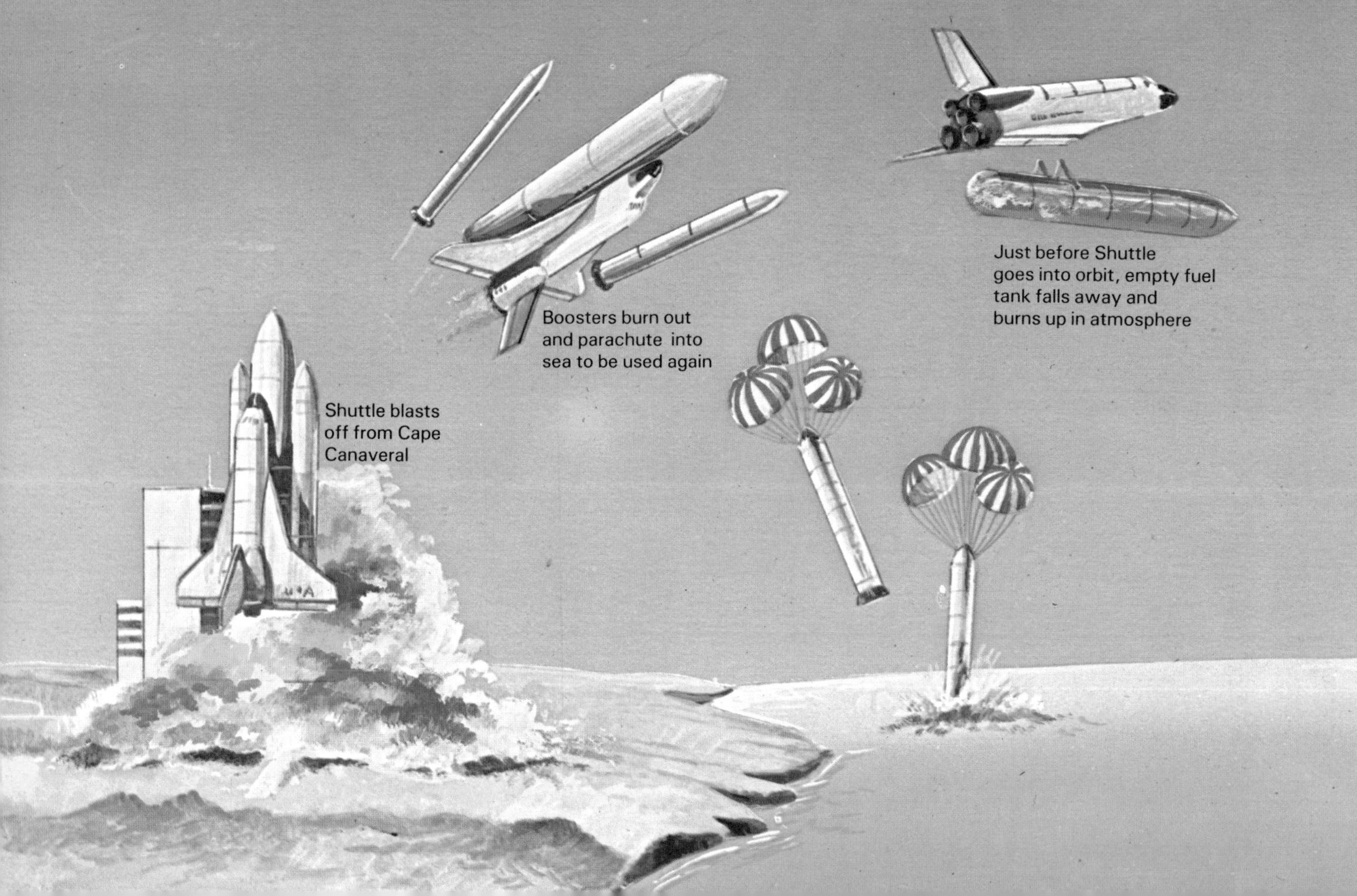

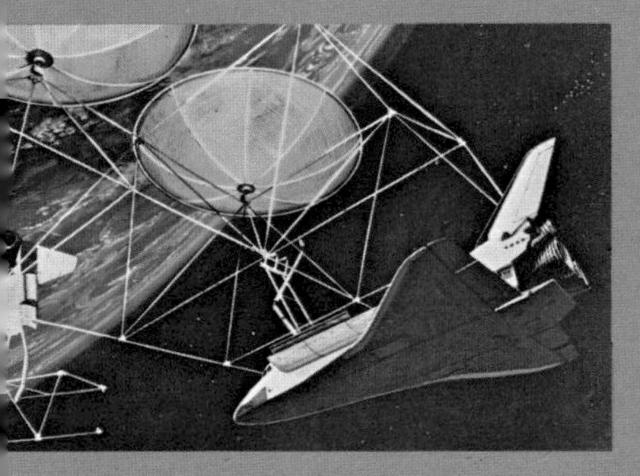

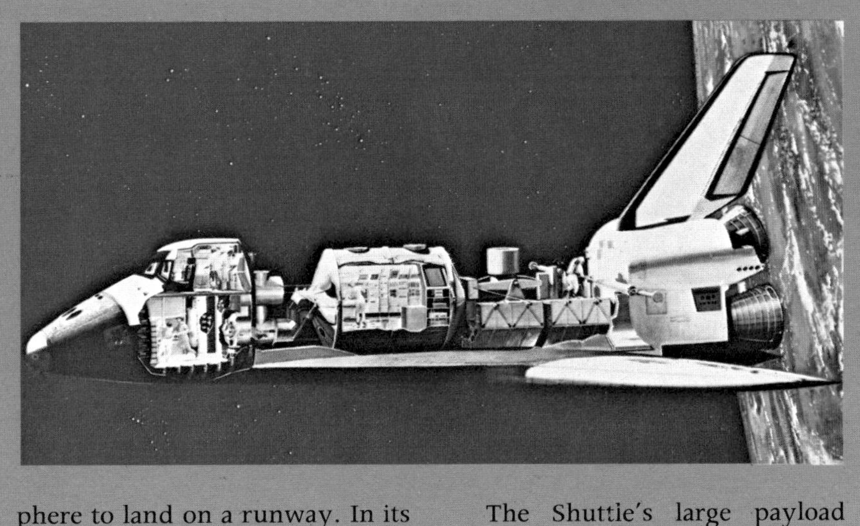

Above: Construction by Shuttle of large antennae in orbit for worldwide communications.
Right: Shuttle with Spacelab space station aboard.

system, this external tank is the only part not scheduled for re-use–despite the fact that, at 33 tonnes, it weighs more than Orbiter's payload.

The Orbiter is scheduled to stay in space for up to a week at a time before returning to Earth. It is covered with insulating tiles which resist the heat of re-entry as it glides back through the atmosphere to land on a runway. In its hangar it can be made ready to fly again within a week. Each Orbiter is expected to be re-used at least 100 times. Test landings of an Orbiter carried aloft by a Jumbo jet were made in 1977, and the first Shuttle flight into orbit is expected in the early 1980s.

The Shuttle's large payload capacity means that it can launch several satellites simultaneously, and it is expected to replace conventional disposable rockets almost completely. However, the Shuttle Orbiter itself will only go into a low orbit around Earth, so satellites destined for high locations, such as geostationary orbit, will have to be manoeuvred there by another rocket stage called the Space Tug. Similarly, probes intended for the Moon and planets will have to be boosted out of orbit by an additional rocket stage.

In orbit

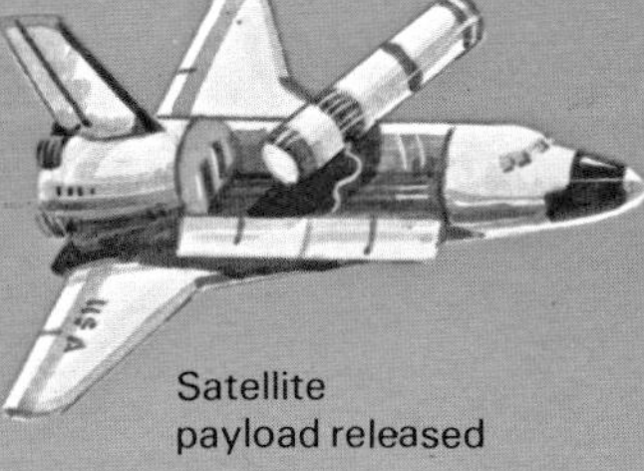

Satellite payload released

Re-enters atmosphere

Lands like a conventional aircraft

Space Shuttle can carry several satellites at a time, and launch or retrieve them by a long arm.

Eventually, the Shuttle may be flying as often as once every few weeks, depending on demand. Many launches will be made by the US Air Force from their Western Test Range in California for military purposes such as reconnaissance. Other operations will be commercial, like those of a cargo aircraft, open to anyone who can pay the freight rates. Civilian Shuttle flights will continue to use NASA's launch facilities at Cape Canaveral.

SKYLAB

Shuttle operations will build on the experience of working and living in space gained by astronauts during the Skylab missions. Skylab was a space station made from the converted third stage of a Saturn V. It was the largest spacecraft ever launched, weighing 75 tonnes and over 25 m (82 ft) long. Its main section was a cylinder 14.7 m (48 ft) long and 6.6 m (21.5 ft) in diameter, divided into a work section and living quarters for three crewmen. Other sections of Skylab included a docking port for access to the station and an airlock for making space walks.

Skylab contained equipment for carrying out numerous experiments such as the processing of materials and observation of Earth, as well as to study the effects of prolonged spaceflight on the human body. At the space station's side was a battery of six telescopes to observe the Sun; these telescopes were operated by the astronauts from a control panel inside Skylab.

The Skylab project followed the Apollo Moon missions. Its purpose was to use as much of the same hardware as Apollo to demonstrate man's ability to live and work for long periods in space. In the event it turned out to be a brilliant success–but only after a disaster that could have wrecked the entire project.

Skylab in Trouble

Skylab was launched on May 14, 1973, by Saturn V. Since the

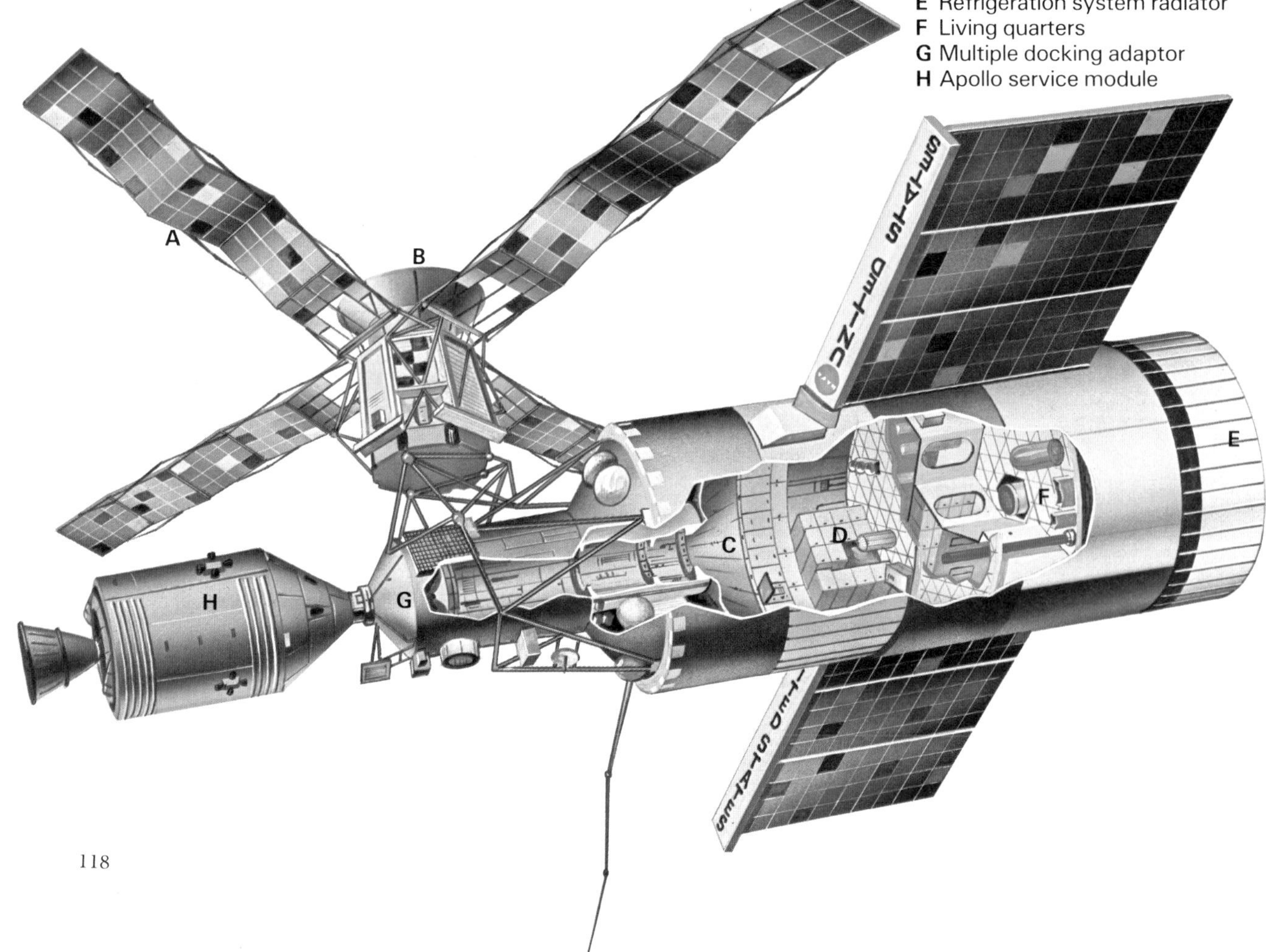

Skylab and Apollo
A Solar cells
B Apollo telescope mount
C Upper experiment compartment
D Equipment storage
E Refrigeration system radiator
F Living quarters
G Multiple docking adaptor
H Apollo service module

space station itself constituted the top stage of the rocket, only the first two stages were 'live'–but these were quite powerful enough to place it into orbit 430 km (267 miles) high.

Trouble struck just over a minute after lift-off. A shield which should have protected Skylab from tiny meteorites while in space tore off as the rocket thundered into the sky, taking with it one of the space station's main power-giving solar panels and jamming another. Once in orbit Skylab began to overheat rapidly because the meteorite shield was designed to act as a sunshield. Skylab was also depleted of power because of the damaged solar panels.

Launch of the first Skylab crew, scheduled for the following day, was cancelled while engineers studied the problem. When the crew took off 11 days later in their Apollo capsule, commanded by veteran astronaut Charles Conrad, they carried with them equipment to repair some of the damage to the space station. They tried without success to free the jammed solar panel before docking with Skylab. Some electrical power to the station was coming from cross-shaped solar panels attached to the telescope mount, but this would not be enough for full operation of Skylab when it was occupied.

Cautiously, the astronauts–Conrad, Paul Weitz, and Joseph Kerwin–entered the stricken space station, finding it hot but not unbearably so. Despite fears that dangerous gases might be emitted from materials inside Skylab as a result of the abnormally high temperatures, the atmosphere proved perfectly safe. The first task for the astronauts was to erect a temporary shield to cut down the Sun's rays hitting the outside of the station. This device was like an umbrella, fed out through a small airlock originally intended to hold scientific instruments. Once the sunshade was installed, temperatures began to drop.

But electrical power was still a problem. Wielding specially prepared metal cutters, astronauts Conrad and Kerwin donned spacesuits and clambered out of Skylab to cut free a metal strap that had been jamming the remaining main solar panel. As this wing-like panel opened to its full extent, electrical power surged into the crippled space station. It had been an impressive example of man's ability to perform difficult and dangerous tasks in space.

Below left and top inset: Skylab in orbit with golden sunshade. *Bottom inset:* Skylab as it appeared after launch, with jammed solar panel and missing micrometeoroid shield.

Urban Spacemen!

After this, the crew settled down to a more normal routine. Skylab had more home comforts than the cramped spacecraft of previous missions. Astronauts slept zipped up in sleeping bags. (In weightlessness, beds are unnecessary.) They ate a wide variety of foods, some in cans and some in bags which could be mixed with hot water to make nourishing soups and stews. Astronauts had special words of praise for the toilet, which in the absence of gravity used a flow of air to draw wastes from the body. There was even a shower, again using a flow of air in place of gravity to create a stream of water droplets.

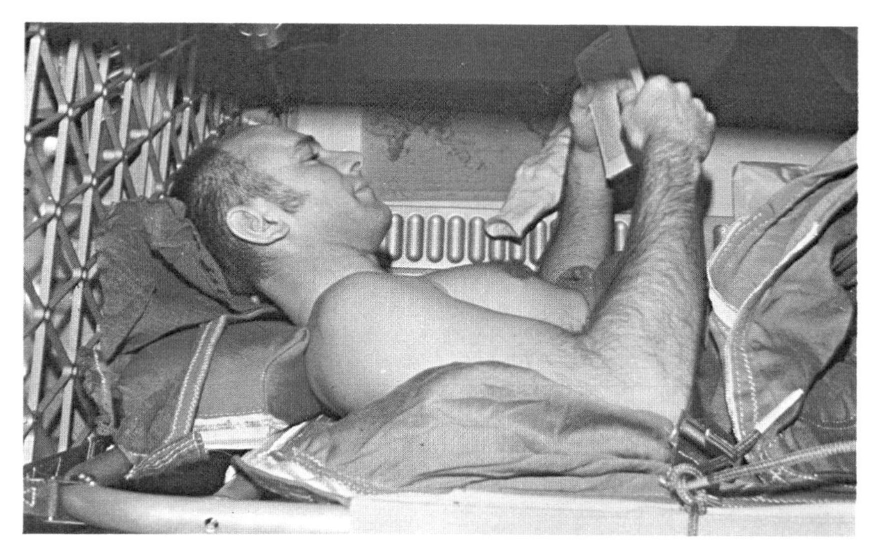

Above: Skylab astronaut Alan Bean reads in his sleeping bag.
Right: Owen Garriott asleep.

Medical Matters

Skylab's first crew stayed in orbit for 28 days, at that time a record. Physical changes occur to the body during prolonged periods of weightlessness. Heart, muscles and bones all weaken, and there are changes in blood cells. The Skylab astronauts kept a check on all these changes, particularly scientist-astronaut Kerwin, a qualified doctor. Among the equipment on Skylab to counteract these changes were an exercise bicycle and a treadmill designed to strengthen body muscles and the heart-lung system. All astronauts on Skylab adapted readily to long-duration spaceflight, and there were no adverse reactions on their return to Earth.

Once the effects on the first Skylab crew of their four weeks in space had been assessed, a second crew–Alan Bean, Jack Lousma, and Owen Garriott–took off for an even longer mission, carrying with them new supplies and experiments. Initially, all three astronauts were affected by motion sickness, but this passed after a few days. Astronauts Garriott and Lousma made a six-and-a-half-hour space walk during which they loaded new film into the solar observatory telescopes and spread out a new and larger sunshade to cool the space station. This shade remained in place for the rest of the Skylab mission.

An interesting experiment performed during this second manned period was selected from suggestions by American high-school students. It involved watching two spiders to see how they managed to spin webs in weightlessness. Arabella, as the first spider was named, had trouble with her initial attempts,

Below: Skylab astronaut Pete Conrad on exercise bicycle.
Right: Jack Lousma sponges down after a shower.

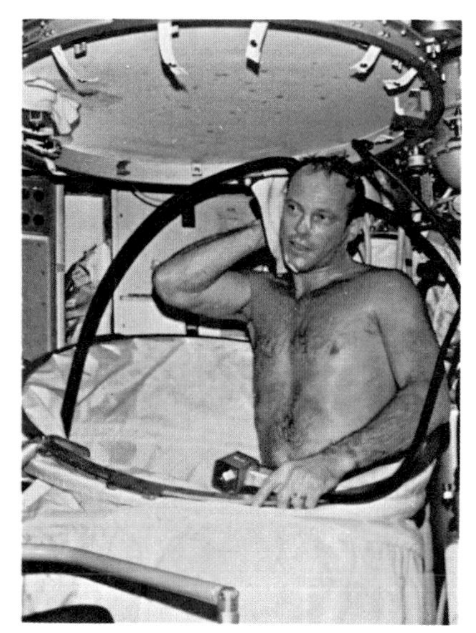

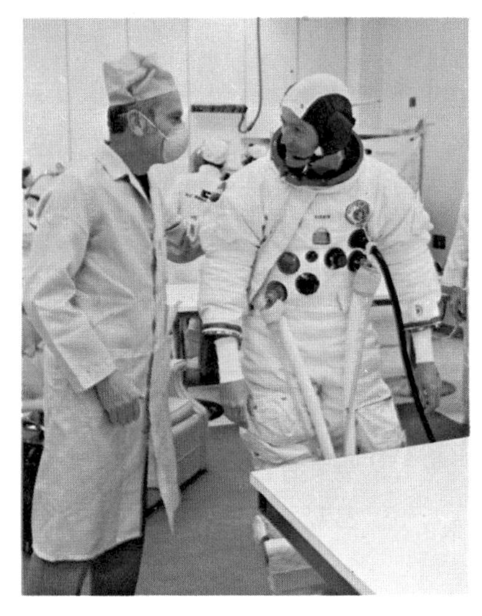

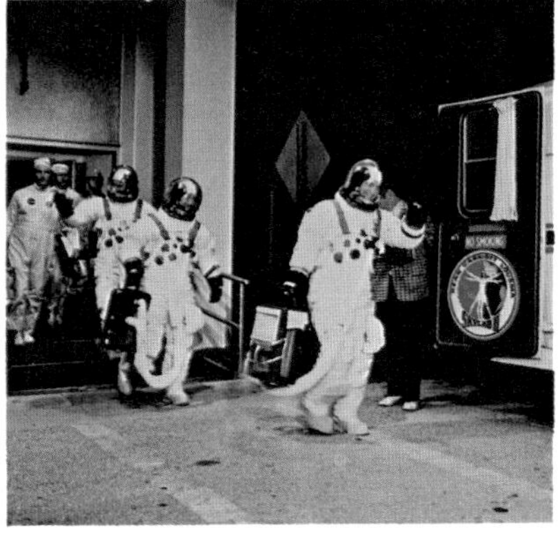

Left: Skylab astronaut Joe Kerwin prepares for launch.
Above: Second Skylab crew of Bean, Garriott and Lousma head for the launch pad.

but managed to spin near-perfect webs after a few tries. Anita, the second spider, had already acclimatized to weightlessness when her turn came, and she performed even better. In another experiment, minnows brought from Earth were observed to be swimming in abnormal tight loops, whereas those hatched from eggs in space swam normally.

The astronauts themselves appeared to adjust to weightlessness after about 40 days, when changes in their bodies levelled off. This second Skylab crew took more exercise than their predecessors, and worked so hard that they continually requested more experiments to perform from ground control. Their success during their 59 days in space meant that the visit of the third and final Skylab crew was extended to 84 days.

Neither Gerald Carr, Edward Gibson, or William Pogue had ever before flown in space, but in this one flight they were to outdo all their predecessors. They, too, felt slightly nauseous at first, but soon adapted.

Gibson and Pogue performed a space walk on their sixth day in space to replenish film in the solar observatory and to correct an electrical fault in a radar antenna. Repairs were carried out during the mission to other pieces of equipment which malfunctioned. This ability of the astronauts to isolate and correct faults as they occurred during all three manned periods added considerably to the overall success of the Skylab mission. A scientific bonus for the third crew was the appearance of comet Kohoutek, a bright comet which rounded the Sun in December 1973.

Skylab Activities

Apart from the medical experiments involving the astronauts themselves, activities aboard Skylab fell into three major areas: solar observation, Earth observation, and materials processing. Astronauts monitored the Sun's activity from a control panel linked to the battery of solar telescopes which observed the Sun in X-ray and ultraviolet wavelengths. They were rewarded by the appearance of several

Below: Installing new film in Skylab's solar cameras during a spacewalk.
Right: Owen Garriott eating in Skylab's wardroom.
Below right: Space spider Arabella spins a web in weightlessness.

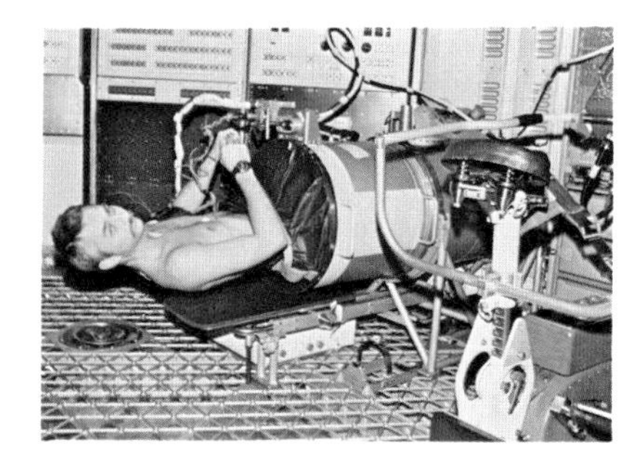

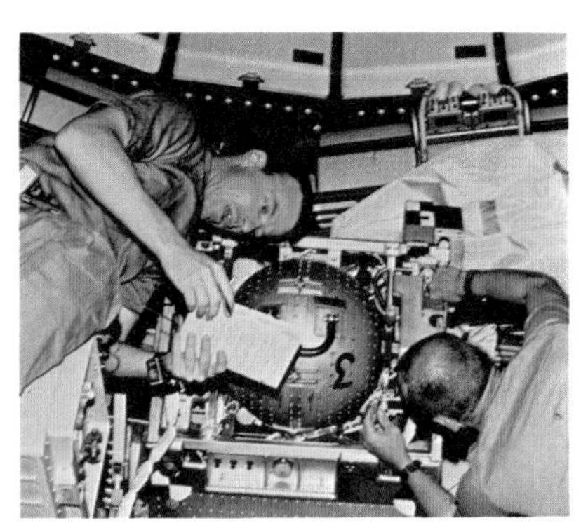

Left: Astronauts Pete Conrad and Paul Weitz check equipment on board Skylab. *Above left:* Owen Garriott undergoes a medical experiment to test his blood pressure. *Above:* Edward Gibson at the controls of the battery of Skylab solar telescopes.

spectacular eruptions on the surface of the Sun which they filmed in detail. Over 180 000 photographs of the Sun were taken by the three Skylab crews, leading to improved understanding of the violent processes operating on our parent star and how they may affect Earth.

Earth itself was a prime target for Skylab's attention. A series of cameras and radar antennae observed Earth at various wavelengths, providing information on a vast range of topics such as soils, crops, forestry, urban developments, water resources, fishing, geological formations, hurricanes, and ocean currents. Over 40 000 Earth photographs were returned from Skylab.

Perhaps most exciting of all were the experiments on processing of materials. In the absence of gravity, materials can be made to blend that would otherwise not mix on Earth. This allows the creation of new alloys which can have greatly improved qualities such as strength combined with lightness. Space processing has another use, too; crystals of so-called semiconductors, glassy materials used in electronics, can be made much more successfully in the absence of gravity. The new and improved materials that can be made in space are likely to be the foundation of important industries on Earth.

Skylab view of Earth, showing Grand Canyon at the left, the Painted Desert, and Lake Powell.

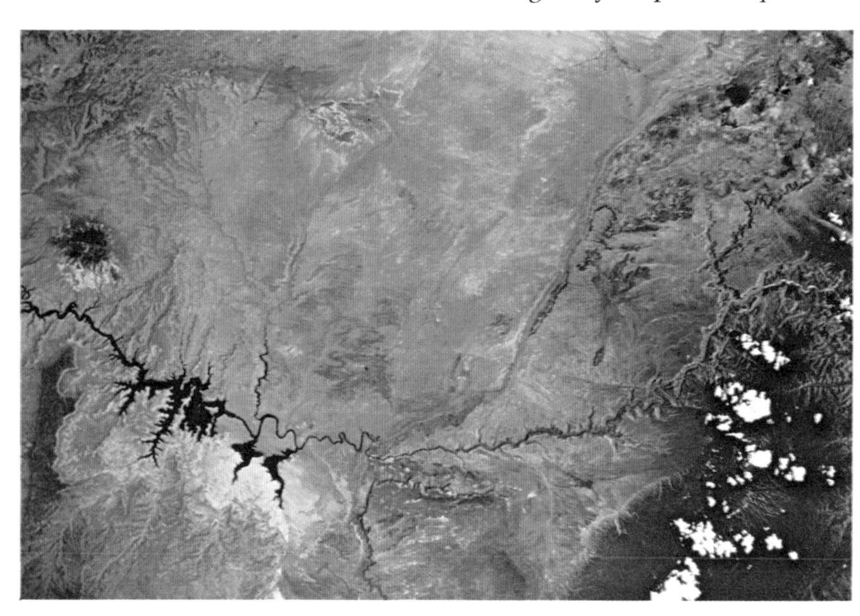

HISTORIC MEETING IN SPACE

When the last crew departed Skylab in February 1974 it was almost the end of the Apollo era. Almost, but not quite. One Apollo mission remained–one with a difference. It was a link-up with a Soviet Soyuz spaceship. This amazing East-West meeting in space had to be carefully planned technically as well as politically, for both spacecraft used different docking mechanisms, and each had its own atmosphere. Soviet craft use normal air at sea-level pressure, whereas American craft use oxygen at low pressure.

To overcome these problems a docking adaptor was built which would allow both spacecraft to link up and which would also act as an airlock between the two. In July 1975, a three-man Apollo spacecraft carrying the docking adaptor blasted into orbit to meet a two-man Soyuz craft commanded by Soviet space-walk veteran Alexei Leonov. Both crews had trained together, as well as learned each others' languages, and the flight was jointly monitored by ground controls in the United States and the Soviet Union. After the successful docking, crew members crawled through the airlock to greet each other and to carry out part of the mission in each others' spacecraft.

After two days linked together the two craft separated and returned to Earth with their own

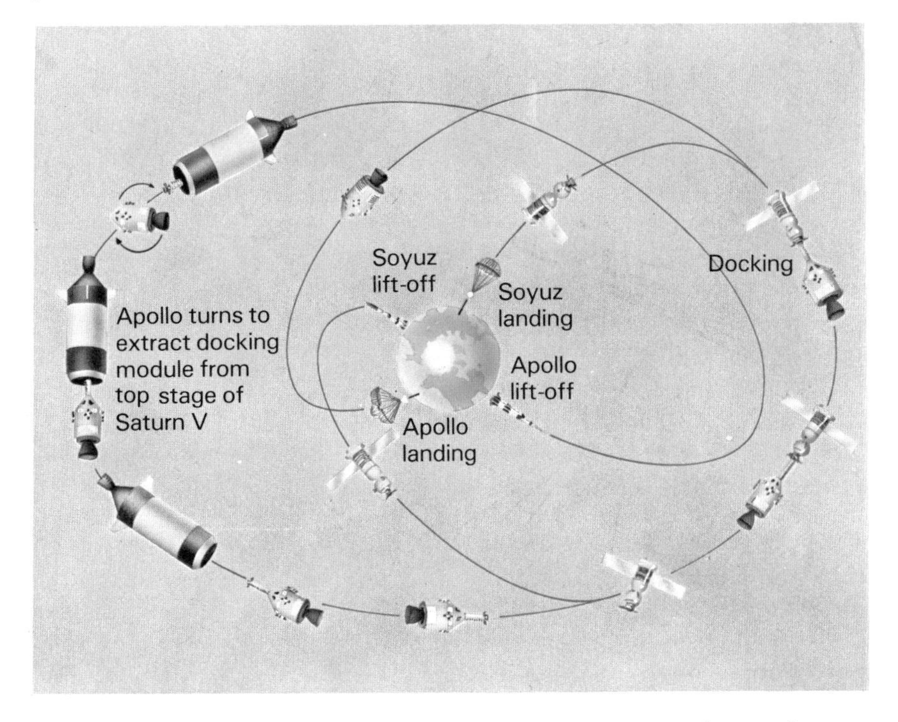

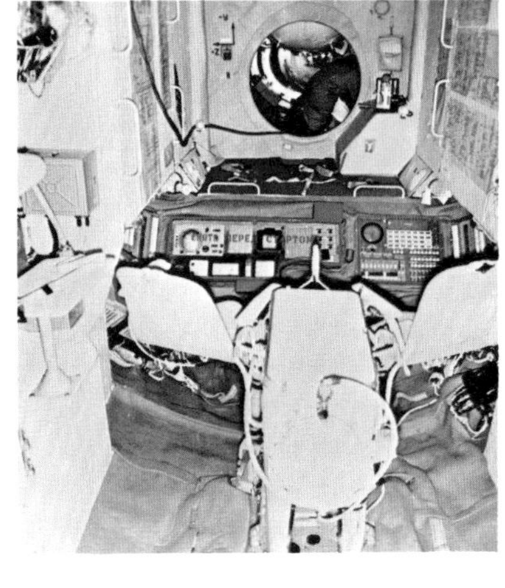

Left: Sequence of events in the Apollo Soyuz joint mission.
Above: The working compartment of a Salyut space station.

crews. The mission's significance was mainly political, leading to improved relations between the space agencies of each country. Further co-operative missions may be undertaken in the future, such as docking the Space Shuttle with a Soviet Salyut space station.

SOVIET SPACE STATION

The Soviet Union has been actively pursuing its own manned space station programme for some years. Its Salyut space station is similar in concept to Skylab, consisting of the converted top stage of a Proton rocket, but

Cosmonauts Alexei Leonov and Valery Kubasov prepare for launch in the Apollo-Soyuz mission.

Salyut and Soyuz
A Soyuz docking radar
B Solar panels
C Experiment working area
D Salyut engine compartment
E Solar panel
F Commander's control desk
G Salyut airlock
H Soyuz working compartment
I Telemetry antenna
J Engine compartment

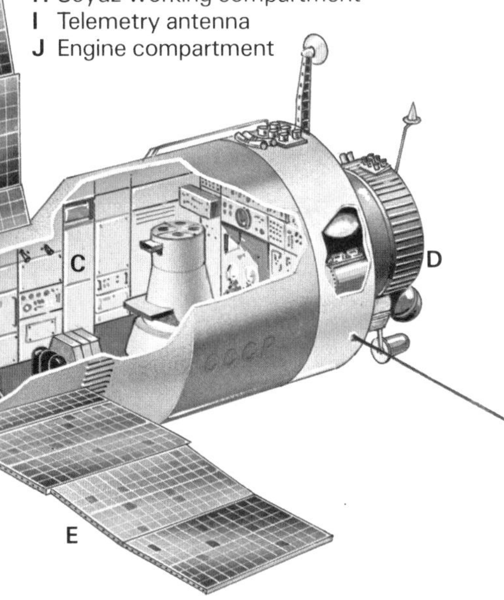

smaller; its overall length is 12 m (39.3 ft) and its maximum diameter 4 m (13 ft), giving it an internal volume about a quarter that of Skylab. Salyut's main aims, like those of Skylab, are studies of human adaptation to weightlessness, materials processing, and observation of the sky and Earth (in this case including some military applications).

Salyut 1 was launched in April 1971, two years before Skylab, but trouble was to dog the space station. An attempted docking by the crew of Soyuz 10 was abandoned, apparently because of problems with the hatch connecting the two spacecraft. The three-man crew of Soyuz 11 subsequently entered the station and completed a 23-day mission in orbit, at that time a record. But when their capsule landed on Earth, all three were found dead in their seats.

At first, it was thought that medical problems caused by the cosmonauts' long stay in space were the cause of their deaths, but the real reason was an accidental loss of cabin pressure during re-entry. They had no defence against depressurization, because Soyuz was too cramped for the cosmonauts to wear spacesuits.

Soyuz, a spacecraft which had already been responsible for the death of one man (Komarov) in its maiden flight (see Chapter 4) was modified again, this time to allow cosmonauts to wear spacesuits during critical phases of their mission. Because of weight and space restrictions, the crew of Soyuz had to be reduced from three to two.

Further trouble struck in 1973 when a second Salyut broke up shortly after reaching orbit. Following test flights of an improved Soyuz, the Soviet space station programme resumed in July 1974 with a 16-day flight in Salyut 3 by Vostok veteran Pavel Popovich, plus Yuri Artyukhin. Theirs was the first completely successful Soviet manned mission in a space station–five months

Artist's impression of Apollo-Soyuz link-up. *Insets, from left:* Soviet Control Centre; Tom Stafford, and Alexei Leonov meet in the docking tunnel; Leonov and Donald Slayton head-to-head; Slayton, Leonov, Stafford and Valery Kubasov congregate at a table in the Soyuz orbital module.

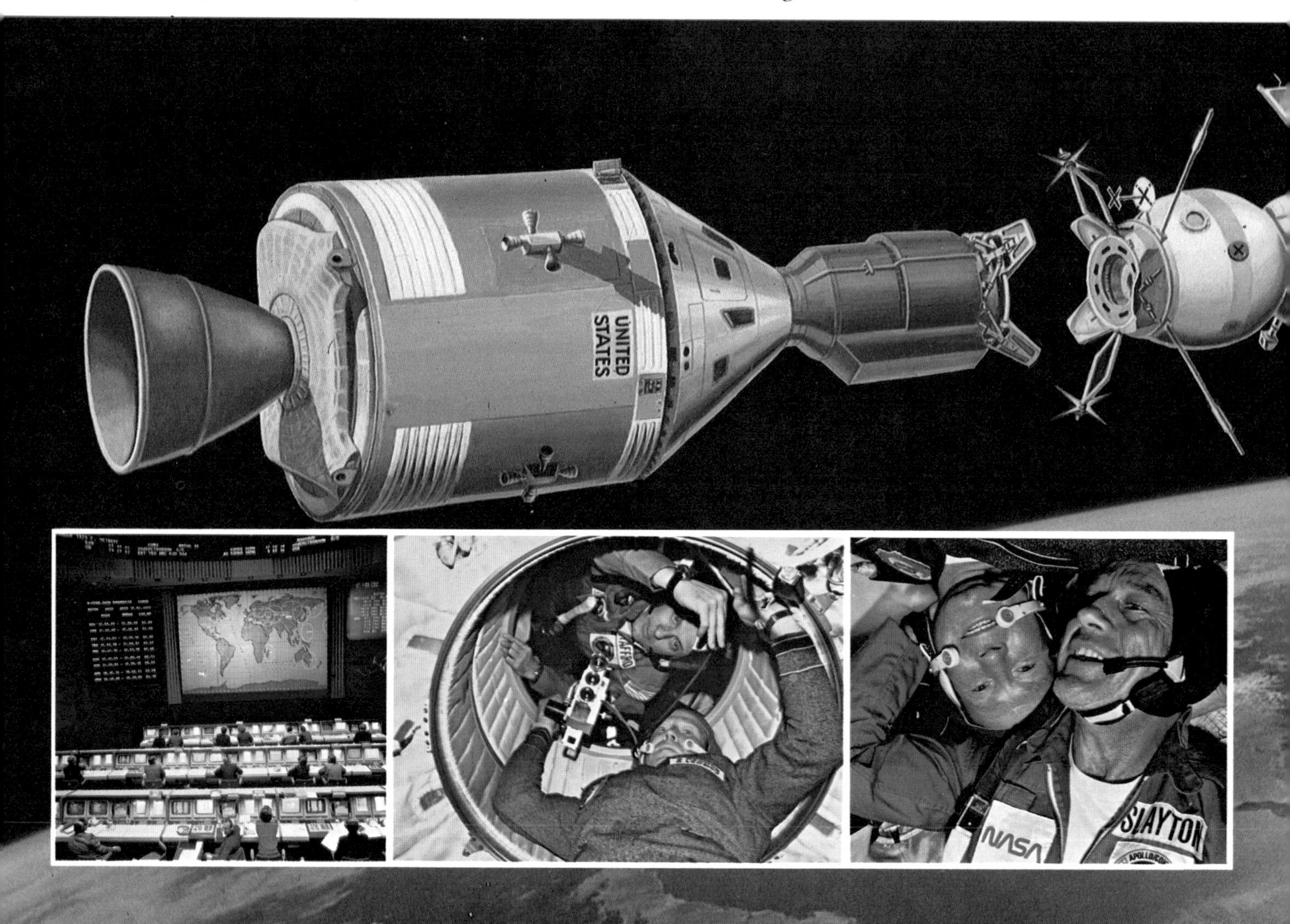

after the return of the final Skylab crew. Salyut 3 was believed to be used mainly for military reconnaissance, as was Salyut 5. Both those stations continued to survey Earth automatically and ejected film capsules for recovery even while unmanned. Salyuts 1, 4 and 6 were mostly scientific in purpose.

With Salyut 4 the Soviet space station programme got into its stride, and began to catch up on the lead established by Skylab. In January 1975 the crew of Soyuz 17 visited the station for a month, and later that year the Soyuz 18 crew of Pyotr Klimuk and Vitaly Sevastyanov completed a 64-day mission in Salyut 4. After they left, the station was resupplied by the unmanned ferry craft Soyuz 20. This was the forerunner of a series of robot tanker vehicles called Progress which ferried up new supplies and equipment to subsequent space stations.

First European in Space

The 64-day flight of Soyuz 18 remained the longest Soviet manned mission until the introduction of Salyut 6, a space station of improved design. It was boarded in December 1977 by the Soyuz 26 crew of Yuri Romanenko and Georgi Grechko, who stayed aboard for a record 96 days–outlasting even the final Skylab crew. During their stay the cosmonauts received week-long visits from two other crews, who brought up extra supplies and returned the results of some experiments to Earth. The second visiting crew contained a special celebrity–a Czechoslovakian cosmonaut, Vladimir Remek, who became the first European (and consequently the first man from outside the US or the USSR) to fly in space.

Once the results of this mission had been digested, the stage was set for an even longer flight–140 days by Vladimir Kovalyonok and Alexander Ivanchenkov, who took off in June 1978.

Boredom is a major drawback when staying many months in space, which is one reason why visitors were sent up—to enliven the cosmonauts' schedule. Among the visitors on this long flight were two more Europeans–a Pole, Miroslaw Hermaszewski, and an East German, Sigmund Jahn.

Several experiments designed by Polish and East German scientists were carried out, including one to find out why food tastes different in space. Spacemen from both East and West had complained that food tastes bland in orbit.

Following the successful return of marathon-men Kovalyonok and Ivanchenkov from their five-month flight, Soviet space doctors speculated that man might be able to stay safely in space for a year or more–long enough for a round trip to Mars. The space endurance record was extended even further in 1979 when Vladimir Lyakhov and Valery Ryumin spent 175 days in Salyut 6.

It has been rumoured that the Soviet Union is developing its own space shuttle to support space station activities. This so-called Kosmolyot space plane is expected to be smaller than the American Shuttle, but it might have a winged first stage like the fly-back boosters originally considered for the American Shuttle.

Left: Cosmonauts Yuri Romanenko and Georgi Grechko. *Below:* Czech cosmonaut Vladimir Remek *(left)*, plus Russian Alexei Gubarev, join Grechko and Romanenko aboard Salyut 6.

SPACELAB

Space station activities in the West will resume in the 1980s with the introduction of Spacelab, a manned laboratory built for NASA by the European Space Agency. Western European scientists will join American astronauts to conduct experiments in orbit. Spacelab will not fly independently like Skylab or Salyut, but will remain in the Shuttle's cargo bay.

Spacelab can come in three versions. The simplest version consists solely of a series of platforms, called pallets, on which instruments can be mounted, open to the vacuum of space. The instruments are operated from a control panel in the Shuttle's pressurized cockpit. In two other versions, either a short or a long pressurized compartment is substituted for some of the pallets, allowing up to four scientists to work in a shirtsleeve environment. Spacelab's total length is 15 m (49 ft) and its width 4.3 m (14 ft).

Spacelab missions are intended to last seven days, but they could be extended to as much as 30 days by addition of extra power sources and supplies. Instruments aboard Spacelab will allow scientists to make observations of Earth and sky, and to carry out experiments in zero gravity. One interesting device is the Space Sled, in which an astronaut will be rocked back and forth in an attempt to learn more about the causes of motion sickness in space.

Each Spacelab is intended to fly about 50 times during a lifetime of 10 years. Once Spacelab is flying regularly aboard the Shuttle, allowing scientists to carry out experiments in zero-gravity processing, the space industrial revolution will get off the ground–literally!

Beyond Spacelab there are ambitious plans for far larger space stations made from components ferried up by the Shuttle. One proposed space station, called the Manned Orbital Facility (MOF), would consist of sections based on Spacelab. At first it would house four astronauts for three months, but could grow by the addition of extra sections to accommodate a total of a dozen or more people.

An alternative idea is to use the main fuel tanks of the Space Shuttle which are discarded once in orbit. One of these, 46.8 m(153.4 ft) long and 8.4 m (27.5 ft) in diameter, could be modified into an expanded version of Skylab, and this too could grow by the addition of extra sections. Space stations like these could become the forerunners of the space colonies discussed in the next chapter.

Large structures such as radio telescopes a kilometre (0.6 of a mile) or more across can be pieced together in orbit by the Shuttle. Framework structures of beams can be made from materials ferried up in the Orbiter's cargo bay. One possible application of structural engineering in space is to build an orbiting power station which beams solar energy to collectors on Earth. Power satellites of this type are being proposed as a solution to our energy crisis.

LATER SOYUZ LAUNCHES

(Continued from page 84)

Mission	*Launch Date*	*Results*
Soyuz 10	April 23, 1971	Vladimir Shatalov, Alexei Yeliseyev and Nikolai Rukavishnikov docked with Salyut 1 but did not enter, possibly because of hatch problem.
Soyuz 11	June 6, 1971	Georgi Dobrovolski, Viktor Patsayev and Vladislav Volkov spent 23 days in Salyut 1, but were killed on re-entry because of pressure loss.
Soyuz 12	September 27, 1973	Vasily Lazarev and Oleg Makarov made test flight of simplified Soyuz for space station ferry missions.
Soyuz 13	December 18, 1973	Pyotr Klimuk and Valentin Lebedev made week-long scientific flight.
Soyuz 14	July 3, 1974	Pavel Popovich and Yuri Artyukhin occupied Salyut 3 for 16 days.
Soyuz 15	August 26, 1974	Gennady Sarafanov and Lev Demin failed in attempts to dock automatically with Salyut 3.
Soyuz 16	December 2, 1974	Anatoly Filipchenko and Nikolai Rukavishnikov rehearsed Apollo-Soyuz mission.
Soyuz 17	January 11, 1975	Alexei Gubarev and Georgi Grechko spent 29 days in Salyut 4.
Soyuz 18	May 24, 1975	Pyotr Klimuk and Vitaly Sevastyanov completed 64-day mission in Salyut 4.
Soyuz 19	July 15, 1975	Alexei Leonov and Valeri Kubasov docked with American Apollo.
Soyuz 20	November 17, 1975	Unmanned refuelling flight to Salyut 4.

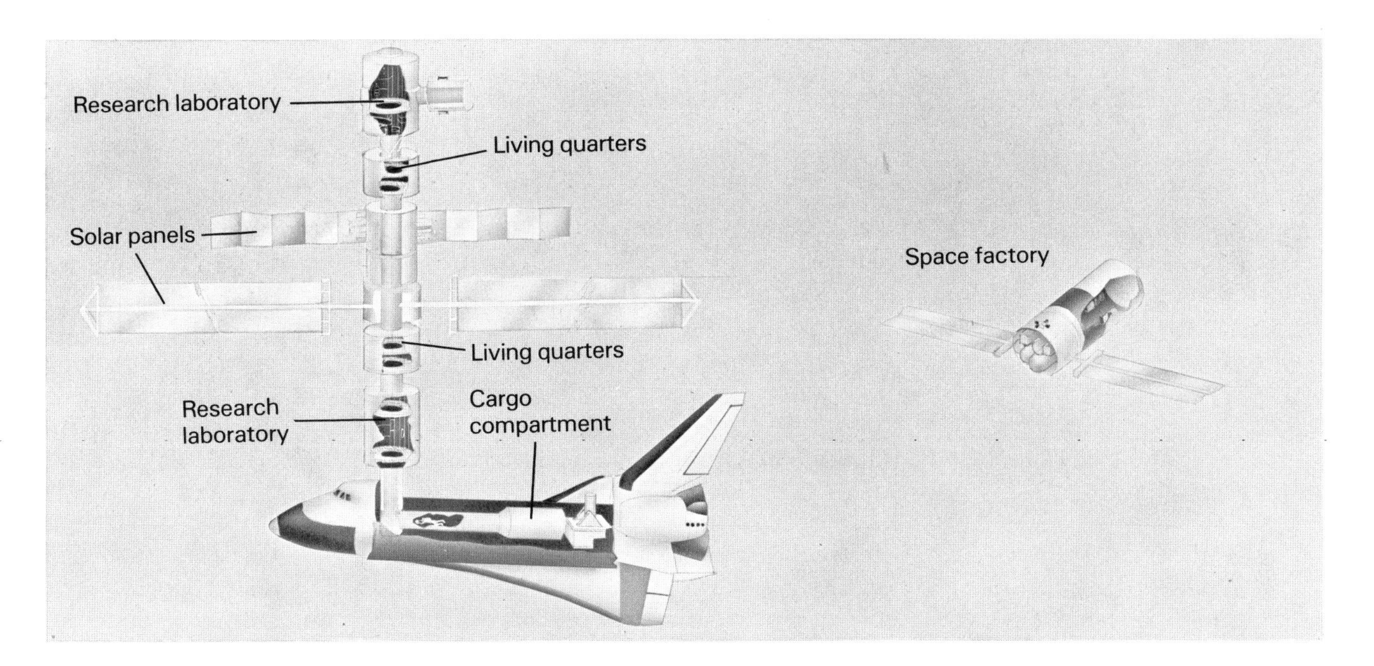

Artist's impression of a possible future space station, made from components ferried up by the Space Shuttle.

Mission	*Launch Date*	*Results*
Soyuz 21	July 6, 1976	Boris Volynov and Vitaly Zholobov spent 48 days in Salyut 5.
Soyuz 22	September 15, 1976	Valery Bykovsky and Vladimir Aksenov made week-long scientific flight.
Soyuz 23	October 14, 1976	Vyacheslav Zudov and Valery Rozhdestvensky failed to dock with Salyut 5.
Soyuz 24	February 7, 1977	Viktor Gorbatko and Yuri Glazkov spent 17 days in Salyut 5.
Soyuz 25	October 9, 1977	Vladimir Kovalyonok and Valery Ryumin failed to dock with Salyut 6.
Soyuz 26	December 10, 1977	Yuri Romanenko and Georgi Grechko made record 96-day flight in Salyut 6.
Soyuz 27	January 10, 1978	Vladimir Dzanibekov and Oleg Makarov visited Salyut 6 for one week.
Soyuz 28	March 2, 1978	Alexei Gubarev and Czechoslovakian Vladimir Remek visited Salyut 6 for one week.
Soyuz 29	June 15, 1978	Vladimir Kovalyonok and Alexander Ivanchenkov made record 140-day flight in Salyut 6.
Soyuz 30	June 27, 1978	Pytor Klimuk and Pole Miroslaw Hermaszewski visited Salyut 6 for one week.
Soyuz 31	August 26, 1978	Valery Bykovsky and East German Sigmund Jahn visited Salyut 6 for one week.
Soyuz 32	February 25, 1979	Vladimir Lyakhov and Valery Ryumin made record 175-day flight in Salyut 6.

Solar Power Satellite

According to the plan, the satellite would collect sunlight on large panels studded with solar cells. The electricity produced would be turned into microwaves (short-wavelength radio waves) with a wavelength of about 10 cm (3.9 ins), and beamed to collector arrays on the ground. There, the microwaves would be converted back into electricity for use on Earth. One such solar power satellite (SPS) in geostationary orbit could deliver 5000 to 10000 megawatts, several times more than the largest power stations on Earth can produce at the moment. A test satellite might be put in orbit in the 1980s, and a full-scale SPS could be in operation by 1996.

In principle, the SPS sounds ideal. Sunlight in space is at least four times as plentiful as at the sunniest spot on Earth, because in space there is no night and no bad weather. Conversion of sunlight into microwaves and back into electricity can be done with high

SKYLAB LAUNCHES

Mission	*Launch Date*	*Results*
Skylab 1	May 14, 1973	Skylab space station, orbited by two-stage Saturn V. World's largest payload.
Skylab 2	May 25, 1973	First Skylab crew, Charles Conrad, Joseph Kerwin and Paul Weitz. Returned June 22.
Skylab 3	July 28, 1973	Second Skylab crew, Alan Bean, Owen Garriott and Jack Lousma. Returned September 25.
Skylab 4	November 16, 1973	Third and final Skylab crew, Gerald Carr, Edward Gibson and William Pogue. Returned February 8, 1974

SALYUT LAUNCHES

Mission	*Launch Date*	*Results*
Salyut 1	April 19, 1971	Occupied by Soyuz 11 crew. Re-entered October 11.
Salyut 2	April 3, 1973	Broke up in orbit.
Salyut 3	June 24, 1974	Occupied by Soyuz 14 crew. Re-entered January 24, 1975.
Salyut 4	December 26, 1974	Occupied by crews of Soyuz 17 and 18. Re-entered February 3, 1977.
Salyut 5	June 22, 1976	Occupied by crews of Soyuz 21 and 24. Re-entered August 8, 1977.
Salyut 6	September 29, 1977	Occupied by crews of Soyuz 26, 27, 28, 29, 30, 31, and 32.

efficiency, so that the advantage of going into space is not lost. Ground collector arrays can be put in remote areas, or even offshore, and the microwave beam will be so spread out as it passes through the atmosphere that it will be far less intense than, say, the beam in a microwave oven. Therefore birds flying through the beam will not be roasted, and aircraft will be protected by their metal skins.

The main drawback is that the SPS will be huge. Each solar panel would measure many kilometres (or miles) on a side, and an entire satellite is estimated to weigh as much as 50 000 tonnes. As the launch capacity of the existing Shuttle is 29 tonnes, to build a full-scale SPS would need daily Shuttle launches for years on end, making Cape Canaveral more like an airport than a spaceport. Exhaust gases from so many rockets could produce severe pollution of the upper atmosphere. Solar power satellites could be ruled out simply because of their weight and size.

Larger and more efficient Shuttles can be built. By replacing the Orbiter of the existing Shuttle with all payload, and adding four improved strap-on boosters in place of the existing two, as much as 150 tonnes could be put into orbit at a time. There are even designs for a super-Shuttle, capable of launching up to 500 tonnes.

Such a vehicle would be expensive and difficult to develop but, once in operation, it would dramatically cut the cost per kilogram (or pound) of payloads into orbit. Only time will tell whether such grandiose vehicles come into existence but, if manufacturing in space becomes a major industry, they may prove an economic necessity.

They would also open up daring schemes of space colonization, as discussed in the next chapter.

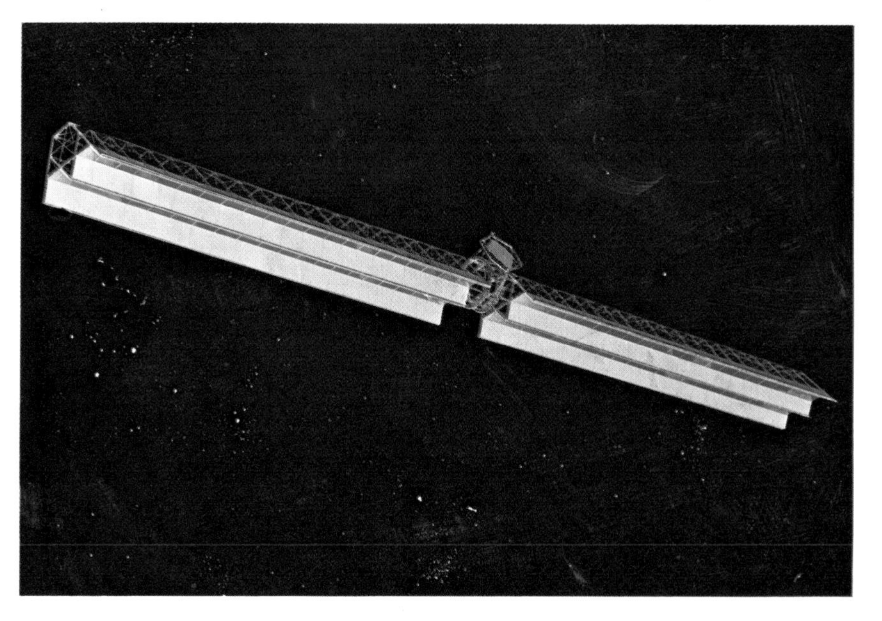

Design for a satellite power station that will collect solar energy in space via solar cells and beam it to Earth in the form of microwaves.

8 COLONIZING THE STARS

Astronauts will one day return to the Moon to set up colonies, and they will also one day colonize the planets, starting with Mars. Further into the future it may be possible to modify the climate of planets such as Mars and Venus to make them more like Earth.

More surprising, perhaps, is the prediction that early next century mankind will begin to colonize the space around Earth in giant habitats built of materials mined from the Moon and asteroids. These self-contained habitats, which would dwarf conventional space stations, would be capable of housing tens or hundreds of thousands of people. Such space colonies, freeing man from the restriction of planetary surfaces, could be the first step in the eventual colonization of star systems throughout the Galaxy.

LIFE ON THE MOON

On the Moon, our first bases are likely to be very simple affairs, similar to the shelters scientists occupy under the ice in Antarctica. The first lunar refuges could be made from empty rocket stages. A few metres (or yards) of lunar soil would insulate them from radiation and meteorites, as well as keeping temperatures constant within the shelter. More ambitious structures made from prefabricated components or even inflatable domes may follow.

Sunlight provides an abundant energy source on the Moon. Oxygen can be extracted from the rocks, where it is abundant. Not only is the oxygen usable for breathing; combining it with hydrogen brought from Earth will produce water. Greenhouses will be established on the Moon in which to grow plants–lunar soil, enriched with suitable nutrients, is a good medium for agriculture.

The Moon has much to offer. Geologists will be keen to work there. So will astronomers, for they can observe the sky without hindrance from an atmosphere. Large telescopes are likely to be set up there, for the Moon

Artist's impression of a possible Moon base of the 1990s, buried underground to protect the scientists from radiation and meteorites.

provides a more stable base than any orbiting satellite. Radio astronomers will want to install their instruments on the Moon's far side, where they will be insulated from Earth's radio chatter by nearly 3500 km (2175 miles) of solid rock. Much larger instruments can be built on the Moon than on Earth because the Moon's lower gravity means that everything there weighs less.

Should our colonists wish, they can make the Moon more like the Earth by introducing an atmosphere. The technique of making a body more Earthlike is referred to as terraforming. Oxygen, released from Moon rocks by heating, would provide a substantial breathable atmosphere which could be replenished as gas atoms leak away. An atmosphere would spoil the Moon's appeal for astronomers and others, but it would give it a habitable climate and screen the surface from dangerous cosmic radiation and meteorites.

TERRAFORMING MARS

Mars is likely to be the next port of call beyond the Moon. A manned Mars trip will most likely be made by two spacecraft, each housing three to six astronauts. Because of the difficulty and expense of such a mission, one craft might be built and launched by Western nations and the other by the Eastern bloc. Should disaster befall one craft during the round-trip of 15 months or more, the other craft would act as a lifeboat to get everyone home. Mars bases, which might be set up well into the next century, would follow the lines of those on the Moon.

If we wanted to, we could employ one of the daring schemes proposed for terraforming Mars. The key lies in the Martian polar caps, which contain plentiful frozen carbon dioxide and water. If we could release these substances, Mars would possess a denser, warmer atmosphere than at present. And the best way to melt the polar caps would be by planting vegetation around their edges. The vegetation, being dark in colour, would absorb sunlight, warming the caps and thereby releasing the frozen gases they contain. An atmosphere of carbon dioxide and water vapour would trap the Sun's heat very efficiently, warming the planet still more and melting its permafrost layer so that liquid water would flow on the surface. The atmosphere would not be breathable, but here again the plants would help by breaking down the carbon dioxide to release oxygen. There is plenty of oxygen in the rocks of Mars and, as the Viking landers found, it is released in large quantities when the soil is dampened. So there is no shortage of suitable gases to form an atmosphere that would make Mars a comfortable place to live.

THE TAMING OF VENUS

Even the hostile atmosphere of Venus might be tamed by the terraformers. In 1961, American astronomer Carl Sagan made the classic terraforming suggestion of scattering the atmosphere of Venus with blue-green algae, primitive plants which would break down the carbon dioxide to release oxygen. As the atmosphere of Venus slowly cleared, temperatures would drop until the conditions became more Earthlike. Venus lacks sufficient water, but this shortcoming could be remedied by diverting icy comets into the clouds, where they would melt to provide rivers and lakes on the terraformed planet.

FREEMAN DYSON'S BIZARRE IDEAS

Of all the ideas for changing the environment of celestial bodies, undoubtedly the most bizarre have come from Freeman Dyson, a

Artist's impression of astronauts exploring the surface of Mars after the planet has been terraformed.

physics professor in Princeton, USA. Professor Dyson has suggested that one day, when biologists are able to tinker suitably with living cells, we might modify trees so that they can grow on comets, thereby creating organic spaceships.

Comets, Dyson says, have all the chemical ingredients necessary for life, such as carbon, nitrogen, water, and minerals, but they lack air and warmth. Dyson's biologically engineered trees would have their roots embedded in the rich material of the comet. The air and warmth produced by their leaves, genetically redesigned to work in a vacuum and specially insulated against ultraviolet light, would be channelled down through the tree trunk to be released inside the comet, making its interior habitable. Seeds from the trees, speculates Dyson, could drift across space to take root on other comets, thereby starting a wave of greening throughout the Galaxy.

Dyson Sphere

Equally fantastic, Dyson foresees that a super-advanced civilization might dismantle a planet such as Jupiter to form a shell of orbiting particles around its parent star. Such a shell, known as a Dyson Sphere, would trap all the parent star's radiation for the civilization to make use of, as well as providing a vast surface area for the civilization to inhabit. From outside, a Dyson sphere would glow a dull red. Several objects like this have been discovered, but they are all believed to be natural shells of dust around stars.

Who can tell whether these fantastic schemes will come to pass? Tampering with the climates of planets, or even dismantling them entirely, is likely to provoke hostile reaction from conservationists. But even if we do not terraform planets or build a Dyson Sphere, someone else may do so.

O'NEILL'S WORLDS IN THE SKY

Much more realistic are proposals for large-scale colonies in space, as championed by the American physicist Gerard O'Neill of Princeton. Some years ago, O'Neill realized that mankind would be better off building his own habitats in space rather than relying on the available land area of the Moon and planets.

We would be better off because the conditions inside a habitat can be tailored to our liking and also because, surprisingly enough, they can provide nearly limitless room for habitation–far more than the surfaces of all the planets and moons in the solar system. We can create our own mini-worlds for living in space, free of the pollution and over-crowding that afflicts so much of our planet today.

These worlds in the sky would be a natural step to follow the large space stations now being considered as bases for manufacturing in orbit. The colonies would be spheres or cylinders, spinning to provide gravity on their inner surfaces.

Inside these sealed habitats,

Professor Freeman Dyson has proposed dismantling a planet to trap energy from the Sun.

A star ship roars past a Dyson Sphere, a shell of orbiting particles housing a super-advanced civilization.

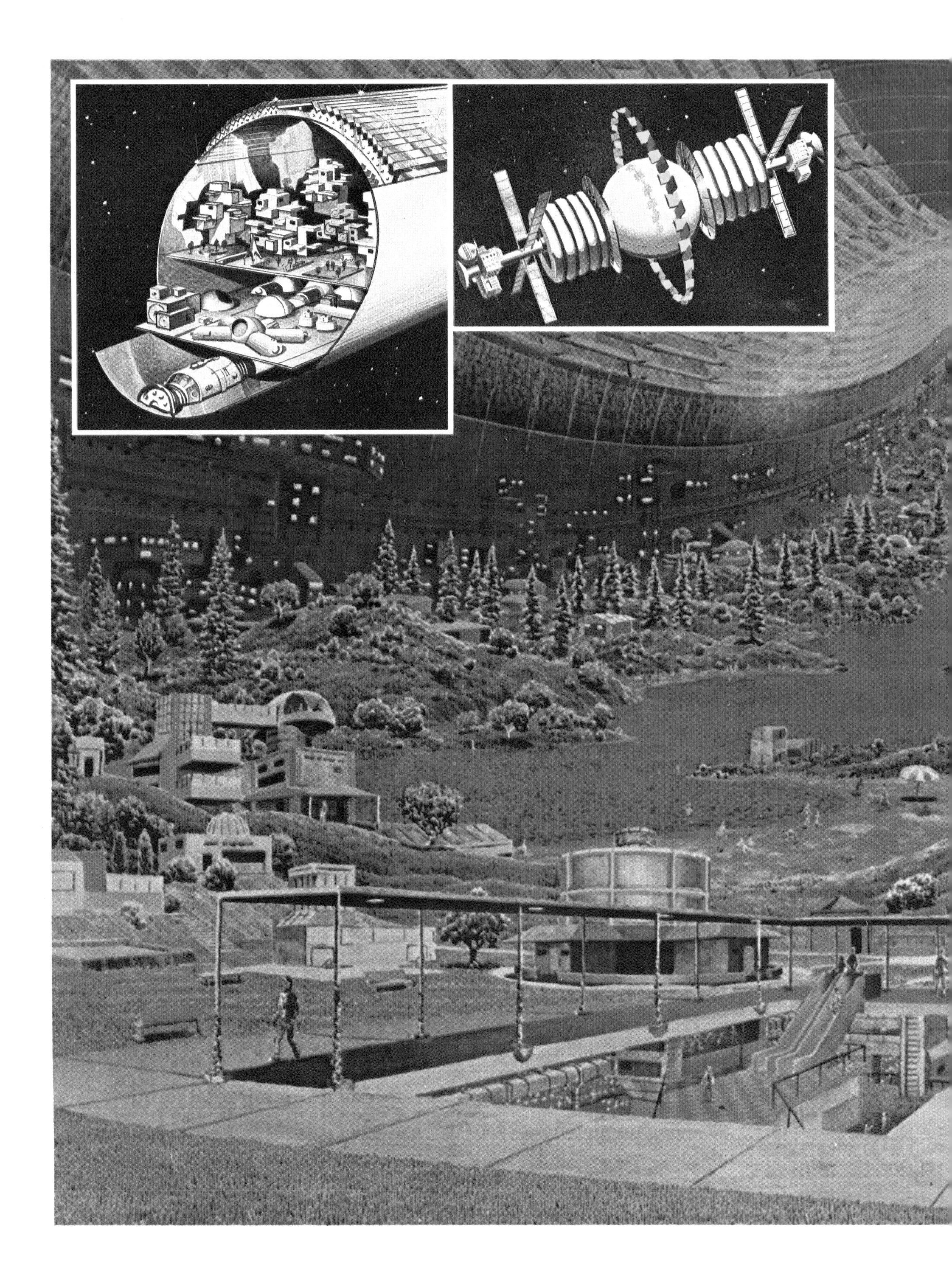

humans would be able to walk around as freely as on a summer's day on Earth. Light and warmth for the colonies would come from the Sun, and solar power would provide their energy needs. They would be landscaped to look as much like Earth as desired, with trees, gardens, parks, lakes, and rivers. Transportation would be by pollution-free electric cars. Life in space, in this vision, could be idyllic.

How do we build such a space world? Ferrying up the materials from Earth would not be possible. So, says O'Neill, we will use the raw materials available in space– in particular, from the Moon. The Apollo missions showed us that the Moon's rocks are rich in metals such as titanium, aluminium and iron, all of which would be good for construction purposes. The surface rocks of the Moon can be scooped up by a small mining team and fired into space.

There are two advantages of launching from the Moon rather than from Earth. Firstly, the Moon's small size means that its gravity is one-sixth that of Earth, so that its escape velocity is lower. And, secondly, it has no atmosphere, so that a high-speed projectile fired from its surface will not be burned up by air resistance. Therefore, we could literally catapult packages of Moonstuff into space, given a sufficiently strong catapult.

Space Railway

O'Neill has designed just such a device to do the job. It is a railway along which buckets containing lunar rock and soil are accelerated by magnetic fields to escape velocity of 2.4 km per second (1.5

Left: Artist's impression of the inside of a wheel-shaped space colony.
Inset, top left: Section through the wheel.
Inset, top right: External view of a sphere-shaped space colony.

Artist's impression of O'Neill's space railway, along which 'buckets' containing lunar rock and soil are accelerated by magnetic fields. These parcels of Moonstuff then fly on out into space.

miles per second). The buckets are then suddenly decelerated and return along rails to be refilled, while the parcels of Moonstuff fly on out into space where they are collected in a vast bag. Once this bag is full, it is transported to the space industrial centre where the lunar ores are processed to extract their metal content. Metal sheets and girders are cast, from which the colonies begin to take shape.

Island One

In O'Neill's design, the first colony, called Island One, is a sphere 460 m (503 yds) in diameter, spinning twice a minute to provide Earth-normal gravity at its equator. Within this sphere, up to 10 000 people could live.

The view would be strange. The sphere would curve around you on all sides, and people on the opposite side of the sphere would have their heads pointed towards you. As you walked towards the axis of the sphere, the gravitational forces caused by rotation would drop; at the axis itself, there would be no rotational force, and hence no artificial gravity. Here, pursuits such as man-powered flight would be possible, or even a zero-gravity swimming pool. Climate inside the colony, as well as night and day, would be controlled by the amount of sunlight admitted to the sphere.

Home from Home

The interior of the colony would be lined with Moon soil, where plants could grow. Houses would be made of brick and glass, again using the raw materials of the Moon. Slag left over from industrial processing will coat the exterior of the colony to protect it from dangerous radiation and meteorite impacts. Agriculture would be carried out in giant greenhouses outside the cylinder. Sunlight will be plentiful in space and crops should flourish, making the colony completely independent of Earth. All this is within the technological capability of the next 30 years, according to O'Neill, who predicts that the first colonies could be dotting the night sky like bright stars early next century.

Many industries will move out into space next century, where raw materials from the Moon and asteroids and energy from sunlight are more plentiful than on Earth. Work forces for these space industries will be housed in the O'Neill colonies. One of the main tasks of the colonies may be to build the giant solar power satellites described in the previous chapter, which beam solar energy to Earth in the form of microwaves for conversion into electricity. Because of their size and weight it would be difficult and expensive to construct power satellites from parts launched from Earth. But using materials mined from the Moon sidesteps the whole problem of launch costs. O'Neill sees the solar power satellite as the economic justification for the space colonization project, for revenue gained from selling electric power to Earth would provide the funding to build more, and larger, colonies.

Island Two

The second generation of colonies, which O'Neill terms Island Two, could be cylinders 3.2 km (2 miles) long and 640 m (700 yds) in diameter, capable of housing 100 000 people. These cylinders would be yoked together in pairs for stability, their long axes pointed towards the Sun. Each cylinder would be made in alternating strips of land and window. Each day, mirrors would peel back from the windows to reflect sunlight into the colony. Around the colonies would be smaller cylinders for agriculture to support the population.

At their largest, the colonies could grow to ten times the dimensions of the Island Two cylinders, capable of housing millions of people. A colony that big could act as an overspill for our own crowded planet, providing a new continent for adventurous settlers. Such self-contained habitats also mean that there will always be a surviving pocket of humanity no matter what disasters befall us on Earth. By colonizing space, we're making man as a species immortal.

Two large O'Neill space colonies of cylindrical shape. Each cylinder would rotate around its axis to create Earth-like gravity on its inner surface. Large mirrors reflect sunlight into the colony. The small cylinders arranged in a ring are agricultural areas.

USA

Space colonies need not be confined to the near neighbourhood of Earth or the Moon, although the first ones undoubtedly will be. With suitably large mirrors to collect sunlight, they could exist as far away from the Sun as the orbit of Pluto. Many of them may concentrate in the asteroid belt, which is an even richer source of raw materials than the Moon, particularly for hydrocarbons to make plastics.

REACHING FOR THE STARS

As man begins to colonize space, his interest will turn further afield, away from the solar system and towards the stars. Already there are four probes in

Many strange objects have already found a home in space and doubtless the future will see more activity in our space world.

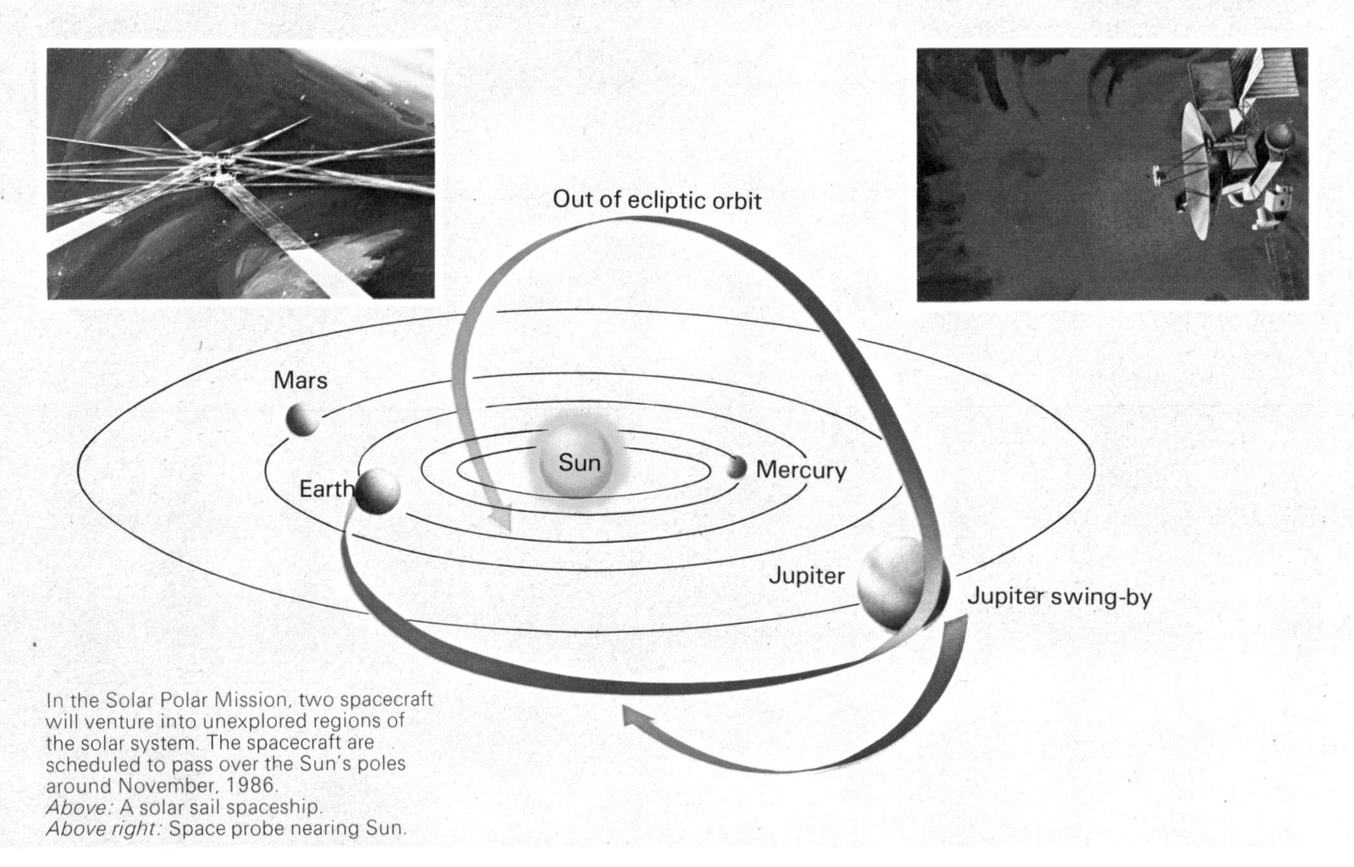

In the Solar Polar Mission, two spacecraft will venture into unexplored regions of the solar system. The spacecraft are scheduled to pass over the Sun's poles around November, 1986.
Above: A solar sail spaceship.
Above right: Space probe nearing Sun.

Left: Artist's impression of an electric rocket, which would eject a high-speed stream of atomic particles to produce thrust.

space which will eventually leave the solar system entirely and drift among the stars. These are Pioneer 10 and 11 and the two Voyager probes, all of which carry messages to any civilization that may one day intercept them (see next chapter). But, even though these are the fastest moving of any probes we have so far sent into space, they would take the best part of 100 000 years to reach the nearest star.

To make interstellar travel an attractive proposition we must go much faster than this. Existing chemical-propelled rockets are large and heavy and, although they are good for providing the initial kick to break clear of Earth, they would need to be of impossibly large size to build up the high final speeds necessary for interstellar travel. Some new form of propulsion is required if we are to cut down the transit time to the stars.

There are various alternatives to chemical rockets, several of which are likely to see service in the near future, at least for travel within the solar system. One is the electric rocket, which ejects a high-speed stream of atomic particles to produce thrust. In an electric rocket, atoms of a heavy element such as mercury or caesium have their electrons stripped away by heat or electric fields. Such a process is called ionization, which is why this type of propulsion is often known as an ion drive. Electric fields accelerate the ionized particles out of the rocket at high speed. Although its thrust is low, an ion drive can continue operating for a long time. Therefore the rocket builds up to a much higher final speed than chemical rockets with a shorter burning time. Small electric rockets have already been tested in space, and they are likely to be employed for space-probe missions to comets or the outer planets.

In the solar system, such rockets will produce electric power from sunlight, but that will not be possible in the darkness of interstellar space. For missions to the stars, electric rockets will need atomic power plants. Atomic-powered electric rockets might reach the nearest stars in a few centuries–considerably better than present capabilities, but still not the ultimate attainable.

Light on the Matter

There is a way in which we can leave the power source of a spaceship behind us, and that is by using the propulsive effect of light. A beam of light exerts a small but distinct pressure, as evidenced by the tails of dust particles that have been wafted away by the energy of sunlight. In the future we may have craft which 'sail' around the solar system propelled by the pressure of sunlight.

There are plans for space probes to rendezvous with comets using just such a system. Lasers can provide an even more intense beam than sunlight, so craft with large, shiny sails that reflect the light of powerful lasers may, in years to come, be roaming the solar system. But it is hard to believe that such a system will be used to reach the stars, although it is possible in theory.

Another suggested craft that need carry no fuel is the interstellar ramjet which would scoop in hydrogen from the gas clouds of space and process it in a nuclear reactor to produce thrust. There are strong doubts that such a device would ever work successfully. An equally fanciful notion is the photon-powered spaceship, which annihilates matter and anti-matter to produce energy for propulsion. This should work in theory, but even so we might never be able to build one because of the exotic engineering problems involved in storing the fuel safely and generating thrust from it.

The Nuclear Pulse Rocket

Of all suggested propulsion systems for interstellar travel, the most promising is the nuclear pulse rocket, which uses the energy of controlled nuclear explosions to kick itself along. Nuclear fusion, the reaction that powers the Sun and stars, offers us an abundant source of energy, one which engineers are attempting to tame in fusion reactors for generating power here on Earth.

Fusion works by squeezing together atoms of a light element to make a heavier one; energy is released as a by-product of the process. In the Sun, atoms of hydrogen are fused together to make helium, a process that also occurs in a hydrogen bomb. By contrast, existing atomic reactors, like atomic bombs, get their energy from splitting heavy atoms, but the energy released is not as great as from fusion. There seems little doubt that fusion is the best energy source to get us to the stars. But how can we put it to use?

A detailed design for a fusion-powered starship has been prepared by a study group of the British Interplanetary Society. They selected as their target Barnard's star, six light years away, because astronomers think it has planets, but the probe could equally well be sent to any other nearby star. The probe, called Daedalus, uses technology likely to be available by the year 2000, so a craft like this could, in theory, be dispatched to the stars next century. And it is expected

A simple diagram to show how fusion works.

Atoms of two light elements are squeezed together

The atoms fuse to form an atom of a heavier element, releasing energy as they do so.

to do the trip in 50 years—less than a human lifetime. It will be unmanned, for this is only a one-way fly-by trip; all operations on board will be controlled by a large computer. But Daedalus engines could be used for subsequent manned voyages to the stars.

DAEDALUS—THE FIRST STARSHIP

Daedalus is a two-stage craft of impressive dimensions–230 metres (251.5 yds) long and weighing over 50 000 tonnes fully fuelled. That 'fuel' consists of billions of tiny, ready-made hydrogen bombs which are fired at high speed into a hemispherical reaction chamber 100 metres (109 yds) in diameter. Here, they are hit by powerful beams of electrons from guns ranged around the chamber's rim. These beams compress and heat the bombs, causing them to explode with a force equal to 90 tons of TNT.

A strong magnetic field lining the reaction chamber catches the force of the blast, ejecting the reaction products out the back and giving Daedalus a little push forward like a spring. The chamber is then ready for another bomb. All this happens very quickly; bombs are injected into the chamber 250 times a second, so the propulsion is virtually continuous rather than a series of jerks.

The first stage is under propulsion for two years. Then it drops away and the second stage takes over. This works in similar fashion, except that its reaction chamber is only 40 metres (131.2 ft) wide. Nearly four years after the probe left our solar system, the second stage burns out and Daedalus coasts for the rest of its 50-year trip to its target at its top speed of 130 million kph (80.7 million mph)–over 12 per cent the speed of light.

Chief designer Alan Bond with the British Interplanetary Society's proposed two-stage starship, Daedalus.

At its head, Daedalus carries a vast payload of 450 tonnes, the weight of five or six Skylab space stations. Into this space are packed about 20 smaller probes to examine interstellar space en route as well as the Barnard's star system itself. As Daedalus approaches its target, the on-board computer dispatches this flotilla of sub-probes for a close-up look at Barnard's star and its planets. Since there is no deceleration, Daedalus and its sub-probes fly past Barnard's star in a matter of hours–a preciously short time after such a long journey.

Each sub-probe radios its readings to the Daedalus mother craft, which relays the data to waiting radio telescopes on Earth. After six years for the signals to cross the six light years to Earth, astronomers will see the first pictures of the face of another star, possibly showing spots, flares, and prominences like our own Sun. According to present astronomical evidence, Barnard's star has giant gaseous planets like Jupiter and Saturn. Will the pictures from Daedalus show these planets to have the equivalent of a red spot and rings like our own Jupiter and Saturn? Perhaps there are smaller planets around Barnard's star, undetected at present, similar to Venus, Mars and Earth. Our first close-up view of another planetary system will help us understand more about our own solar system and its origins. The results could be with us next century.

Daedalus will not be built on Earth. Constructing a starship, or a whole fleet of them, is a job for the space colonies discussed earlier. One problem is finding sufficient fuel to make the bombs. The Daedalus bombs are intended to be a mixture of deuterium, which is a heavy isotope of hydrogen, and helium-3. There is not enough helium-3 on Earth for even one Daedalus, but it is plentiful in the gaseous atmosphere of Jupiter, where it can be sifted to produce sufficient fuel for billions of Daedalus craft. Starships, it seems, will be constructed only by those civilizations with industries spread throughout their own solar system–which, next century, will include ourselves.

No one can tell whether our first starship will be like Daedalus. Perhaps the reality will be even better than our current paper speculations, as the Saturn V far exceeded the dreams of the first rocket designers. The important point is that already we can see how an interstellar mission might be mounted. Once interstellar flight has been shown to be possible in theory, human ingenuity will do the rest.

Will Man Follow?

Daedalus, or whatever our first starship is eventually called, will be only an unmanned pathfinder probe. Will humans follow, as they followed the robot explorers to the Moon and seem destined to follow to the planets? Clearly, the space between stars is so great that interstellar missions must always take longer than journeys to the planets (unless some as-yet-unknown way of skipping across the Universe in the blink of an eye is discovered by physicists). Therefore we must be resigned to journeys lasting several tens of years, even allowing for technical improvements over Daedalus.

Of course, we may never find a way of travelling over interstellar distances faster than Daedalus– nor may we need to. If we are willing to take a slow cruise to the stars, all the items necessary for manned interstellar flight are within our grasp. By attaching a Daedalus-type propulsion unit to an O'Neill-type space colony we could produce that science fiction dream – an interstellar ark.

Sub-probes are ejected from the top stage of Daedalus during the journey to Barnard's star.

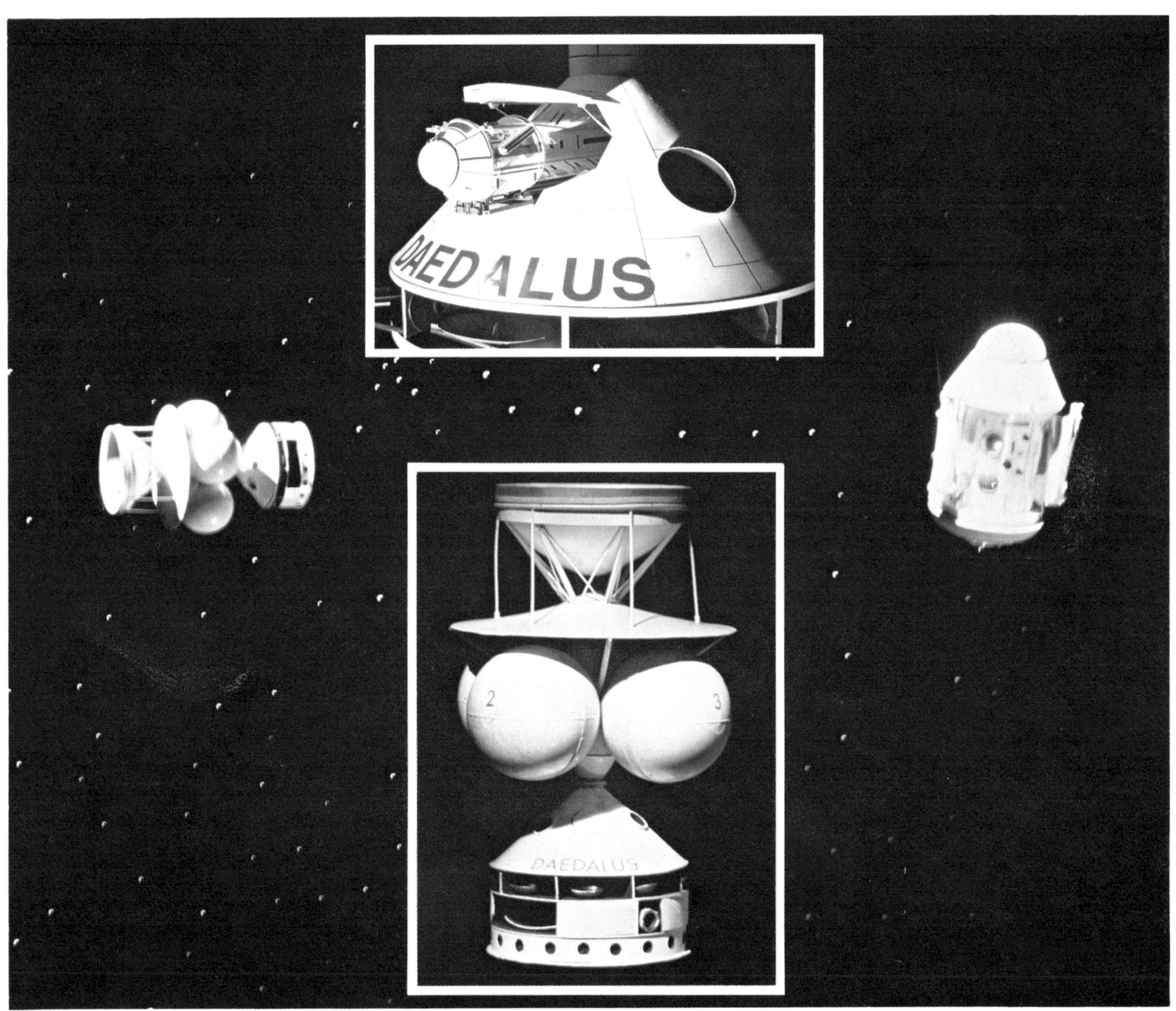

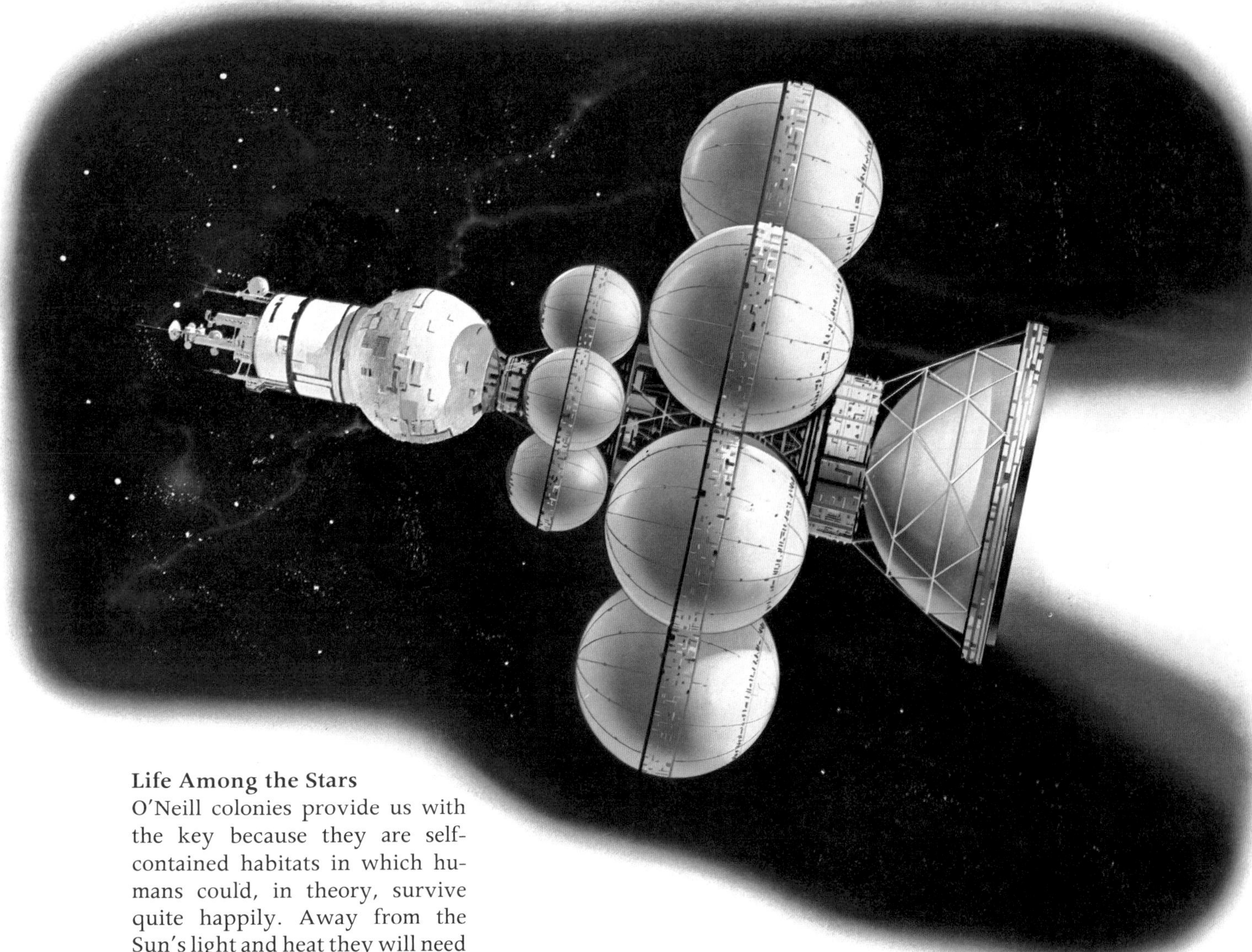

Artist's impression of a possible interstellar ark, in which a colony of humans is propelled to the stars by a nuclear rocket.

Life Among the Stars

O'Neill colonies provide us with the key because they are self-contained habitats in which humans could, in theory, survive quite happily. Away from the Sun's light and heat they will need on-board power supplies in the form of nuclear generators. In a space ark, generations could live, reproduce and die, knowing no other home. The distant descendants of those who began the voyage would step out at the destination, either to begin a new life on planets there or to build more space colonies from whatever raw materials were available.

A trip between stars in such an ark might take a few centuries, depending on the power of the propulsion unit. If there is no better way of sending people on interstellar journeys, then this is how it will have to be done. Not everyone will find the prospect appealing, but there are bound to be a few adventurous people who would be willing to try it. Perhaps there are survey parties from other civilizations roaming in their arks between the stars at this very moment.

Once the inhabitants of an ark have set up base around a new star, they can then begin to explore ahead with robot Daedalus-type probes. Based on the results of these probes more arks may be dispatched, thus extending the spread of humanity (or some alien civilization) into the Galaxy. Eventually, the entire Galaxy may be colonized in this way.

Even at the relatively slow speed of an interstellar ark, and allowing for scouting ahead before each new step, colonization of all suitable stars in the Galaxy would take a few million years–long in terms of human lifetime, but only a fraction of the lifetimes of stars and planets, which are measured in thousands of millions of years. If other advanced civilizations arose long before us around another star and began to colonize space, they would by now be expected to have spread throughout the Galaxy. Therefore, we might ask: Where are they?

At present, it must be admitted that there is no sign of them. Perhaps the whole Galaxy is ours for the taking–or perhaps we have overlooked something.

This brings us to the exciting question of whether or not we are alone in space, which is discussed in the next chapter.

9 LIFE OUT THERE

Life ought to be abundant throughout space. As far as astronomers can tell, there are plentiful planets around other stars to form potential homes for life. Also the chemicals which make up living things exist throughout the Galaxy in the clouds of matter from which stars and planets are born. What we do not know as yet is whether other life has really come into being elsewhere, and if so in what form.

The search for other life in space is one of the most exciting quests in modern science.

SEARCHING FOR OTHER LIFE

Mankind has two methods of searching for other life in space: sending space probes to the planets of our own solar system, and listening with radio telescopes for possible messages from distant civilizations.

Astronomers estimate that there may be as many as a million civilizations, either like ourselves or more advanced, throughout the Galaxy. So it is certainly worthwhile making the effort to find them. Let us now examine in more detail some of the scientific reasons for belief in the existence of life in space, and the ways in which we are looking for it.

Star-gazing

We know that the Sun is an average star. It is an ideal parent star for life-bearing planets, for two main reasons. The first reason is that it has maintained a stable output of light and heat for billions of years, long enough to allow the origin and evolution of living things. Secondly, it gives out enough energy to keep a reasonably large volume of space around it warm enough for life, unlike fainter stars.

Specifically, the Sun is believed to have existed for about 5000 million years, and we can look forward to a similar length of time before it burns out. Smaller, cooler stars live longer than the Sun but they give out less light and heat, so that a planet at Earth's distance would be too cold for life. Larger, hotter stars warm up a larger volume of space, but this apparent advantage is counteracted by the fact that they burn out much more quickly than the

Stars are forming from this gas cloud, called the Lagoon nebula. Do some of these stars have planets bearing life?

Sun, giving insufficient time for life to develop. Had the Sun been twice its actual mass, for instance, it would have burned out after 1000 million years, and we would not be here. Therefore, astronomers and biologists conclude that if we want to find life in space we should look in the vicinity of stars like the Sun.

How many stars actually have planets? At the moment we cannot be sure, because no planets of other stars have yet been seen; they are too faint. Theory says that our own solar system formed from a disc of dust and gas orbiting the Sun after its birth. Therefore, the birth of the Sun and planets was closely linked. Since all stars are believed to form in a similar way, as described in Chapter 1, there is every reason to believe that planetary systems could be commonplace.

In many cases, stars have come into being not on their own but as members of twin or multiple-star systems. Astronomers have found that about four stars out of every five are members of twin or multiple systems. The remaining 20 per cent of stars are apparently single, like our Sun. If the theory is right, such stars should be expected to have planets.

Barnard's Star

There are indirect ways in which the existence of planets around stars can be inferred. One way is to watch for a slight wobble in a star's position in the sky, which might be caused as it moves around its common centre of gravity with a planetary system, like one half of an uneven dumb-bell. This method works only for the nearest stars–the wobble is too small to be noticeable further away–but it seems to have borne fruit in the case of Barnard's star, the second closest star to the Sun.

Professor Peter van de Kamp of Sproul Observatory in Pennsylvania, USA, has found from many years of painstaking photographic observations that Barnard's star shows a slight wobble in its position. He interprets this as being due to the existence of planets orbiting the star. He calculates that one planet is about the size of our own Jupiter, orbiting every 12 years, and the other is like Saturn, orbiting every 20 years or so. Other, smaller planets like Earth might also exist, but their effect would be too small to notice.

Needless to say, Barnard's star is the subject of continuing interest to astronomers who are trying to confirm Professor van de Kamp's results. Wobbles of other stars have been suspected, but no cases give as definite an indication of a planetary system as Barnard's star. When large instruments such as the Space Telescope are put into orbit in the 1980s, we should learn more about the existence of planets around stars.

How Life Begins

Given that a small fraction of all stars do indeed have planets, what are the chances that life will actually get under way on any of these planets? This is one of the biggest question marks in our speculations. As yet we do not know how life comes into being, but we have some clues to the first steps, thanks to some remarkable experiments first performed in 1953 by an American chemist, Stanley Miller, of the University of Chicago.

Dr Miller placed in a flask a mixture of the gases believed to have existed in the primitive atmosphere of Earth. These gases included hydrogen, ammonia, methane, and water vapour. Through this acrid mixture he passed an electrical spark to simulate the effect of lightning in

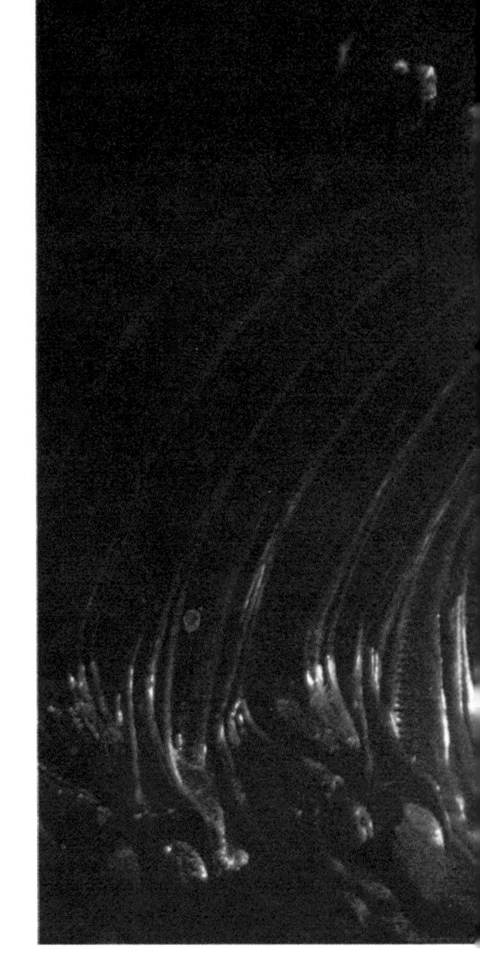

the atmosphere. The energy from the electric spark produced chemical reactions which gave rise to complex organic (carbon-containing) molecules including amino acids, which are the building blocks of life. Miller's experiments have been repeated and extended, confirming that these molecules and others important to life are easily formed under a wide variety of conditions likely to be encountered on planets.

We know that all the right chemicals for life exist in space, because radio astronomers can detect them in the gas clouds of space in which stars are forming, such as the Orion nebula. About 50 chemicals of varying complexity, containing up to 11 atoms, have so far been discovered, and the list is bound to grow. Possibly, amino acids themselves will one day be discovered floating among the stars.

Even more astoundingly, amino acids and other complex organic

Space-suited astronauts explore a mysterious planet in the film *Alien*.

molecules have been found in meteorites, lumps of rock which crash to Earth from space. These molecules must have formed from the cloud that gave rise to the solar system. This finding, allied with the observations of radio astronomers, is impressive evidence that complex organic chemistry can proceed even in the gas clouds of space. When a planetary system comes into existence, it must be surrounded by all the necessary ingredients for life.

According to Professor Sir Fred Hoyle and Professor Chandra Wickramasinghe of University College, Cardiff, Wales, developing planetary systems may be surrounded by more than that. Building on the evidence for organic chemistry in space, they have proposed that life itself originates in gas clouds, and is subsequently scattered on to planets via comets and meteorites.

Most astronomers reject this idea, believing instead that the conditions for the formation of life would be better on planets than in gas clouds, but it does illustrate the extent to which scientists are willing to speculate about the existence of extraterrestrial life. Should Hoyle and Wickramasinghe be right, of course, there may exist some vast, shapeless lifeform in interstellar space, like the famous Black Cloud of Hoyle's science fiction novel.

The mere existence of the right chemicals does not automatically mean that life will actually arise. How the building blocks of life arranged themselves in the right order to form the first primitive cells on Earth is a mystery, and may always remain so. But the process seems to have happened surprisingly quickly, for there is evidence of simple, single-celled organisms like algae on Earth well over 3000 million years ago. Therefore there has been life on Earth for more than three-quarters of its existence, although of course the more advanced forms of animal life did not appear until a few hundred million years ago, and the ancestors of man himself date back only a few million years.

We might get a better idea of how life forms if we could find traces of it on other planets. And the only planets we can reach with space probes at present are those of our own solar system. Mercury and Venus are too hostile for life. Mars long seemed the most promising abode for life elsewhere but, as we saw in Chapter 6, detailed investigations by two Viking space probes have turned up nothing positive.

Jupiter's Atmosphere

Of all planets, Jupiter now seems most likely to tell us more about the possibility of life elsewhere. Jupiter's atmosphere contains the same gases as those from which life is believed to have formed on early Earth; in fact, the atmosphere of Jupiter may be regarded as a fossil relic of the cloud which spawned the solar system. Because of Jupiter's strong gravity it has retained these gases, whereas smaller planets like Earth lost their primeval atmosphere, to be replaced by other gases released from their interiors through volcanoes.

Scientists imitating Jupiter's atmosphere have produced colourful organic molecules that are quite likely to be the cause of many of the colorations in Jupiter's clouds. Jupiter may therefore be a natural laboratory for the steps leading to the origin of life. American astronomer Carl Sagan has suggested that the process may have gone the whole way, and that simple forms of life are to be found floating in the clouds of Jupiter.

On Sagan's speculation, the simplest forms of Jovian organisms would be like plankton in the seas of Earth. Indeed, some terrestrial plankton are coloured red, which is the predominant coloration of Jupiter. Larger creatures might also exist, the equivalents of fish and whales, feeding on the Jovian plankton. These creatures would be like balloons, able to float in levels of the clouds which were wet enough for life, and moving around by expelling gas.

GALACTIC CIVILIZATIONS?

Sagan's charming speculation is, perhaps, somewhat far-fetched. In fact, it seems most likely that Earth is the only life-bearing planet in the solar system. To find life elsewhere in space we must therefore look to the planetary systems of other stars.

One day we may be able to send space probes to other planetary systems, as we saw in the previous chapter. But for the moment we must hope that there exist in the Galaxy some advanced, technical civilizations capable of making their presence known to us over interstellar distances. What are the chances that such advanced civilizations exist?

Astronomers have attempted to estimate the likely number of advanced galactic civilizations by taking into account factors we have been discussing such as the number of stars with planets, the likelihood of life arising there, and so on. Even if life is fairly abundant throughout space, we have no way of knowing what form it may take. Simple vegetation may be the norm, or sea creatures like those that teemed in the ancient oceans of Earth. Only in a fraction of cases can we expect life to evolve to an intelligent, technological level such as ours.

Even so, with a starting point of well over 100 000 million stars in the Galaxy it seems statistically certain that many civilizations should have emerged by now. But not all of them will have survived. The lifetime of a civilization is an important factor in estimating how many aliens there may be out there at present. A civilization may blow itself to bits shortly after reaching a certain level of technology, as we are in danger of doing with our own nuclear arsenal. Other factors such as destruction of its environment by pollution may limit the lifetime of an alien civilization. Or the

Science fiction films are extremely popular today. *Above:* A scene from *Star Wars. Above right: Futureworld. Right: Close Encounters of the Third Kind.*

civilization may simply lack interest in communicating with its fellow beings in space.

Making allowances for all these factors, astronomers have estimated that there may be one million civilizations, like ours or more advanced, throughout the Galaxy. Of course, this result depends entirely on the assumptions made. With different assumptions the number of estimated civilizations can become still greater, or very much less. But in any case it seems that, statistically speaking, we cannot be alone.

Radio Communication

How can we detect the existence of these predicted alien civilizations? Radio communication seems a very good method. Modern radio telescopes would allow us to detect artificial signals coming to us over interstellar distances, and could also be used to send a reply. Some astronomers believe there could already be an intergalactic radio-communication network between civilizations, and that we might one day receive a call inviting us to plug in.

At the National Radio Astronomy Observatory in Green Bank, West Virginia, USA, Dr Frank Drake made the first deliberate attempt to pick up interstellar radio messages in 1960. He turned the observatory's 26-m (85 ft) radio telescope towards the Sun-like stars Tau Ceti and Epsilon Eridani, both being about a dozen light years away, and tuned in at a wavelength of 21 cm.

Drake's listening attempt was whimsically named Project Ozma, after the mythical land of Oz. No alien signals were heard despite three months of listening, but this is no great surprise as even the most optimistic predictions suggest that the nearest civilizations to us will be much further away than Tau Ceti or Epsilon Eridani.

Dr Drake chose to listen at 21 cm because this is the wavelength emitted naturally by hydrogen, the most abundant substance in space. Radio astronomers all over the Galaxy would know of this wavelength and be ready tuned-in to it, so broadcasting at 21 cm would seem a likely strategy for a civilization to attract attention.

Project Ozma was followed by listening attempts in the Soviet Union, and again at Green Bank, this time using new, larger radio dishes with diameters of 43 m (140 ft) and 91 m (300 ft), capable of detecting far weaker signals than the original Ozma search. These efforts go under the name of SETI–the Search for Extraterrestrial Intelligence. (An alternative title is CETI, Communication with Extraterrestrial Intelligence).

The most comprehensive SETI project to date has been undertaken at Green Bank. In this search, more than 600 stars similar to the Sun within 80 light years were surveyed at 21 cm wavelength, but without success. In a project at the Ohio State University Radio Observatory, USA, the entire sky is being repeatedly scanned for possible alien 21-cm emissions. Other SETI projects have been undertaken at other radio observatories. To date, there have been no reported successes.

The 46-metre (150-foot) radio telescope at Algonquin in Canada is used to listen for positive radio messages from extraterrestrial beings.

On Different Wavelengths

Some people would disagree that 21 cm is such a good wavelength for interstellar signalling–one reason being that it is already very noisy with natural emissions from hydrogen. One alternative being pursued in Canada is to listen at the wavelength of 1.35 cm, which is emitted by water molecules. Lifeforms to whom water is vital,

Arecibo, in Puerto Rico, the world's largest radio telescope. The reflector is a metal net.

such as ourselves, would consider this a very significant wavelength, so the argument runs. Arguments in favour of various other wavelengths have also been advanced.

Perhaps there is no 'best' wavelength for interstellar radio communication after all. Instead, there may be a whole range of wavelengths which we should scan. Of particular interest is perhaps the region between the 21-cm line of hydrogen and the 18-cm line emitted by a substance called hydroxyl, chemical formula OH; the attraction of this region is that it coincides with the minimum of background noise from natural sources in space. Since hydroxyl (OH) and hydrogen (H) go together to make water (H_2O), this area of the radio spectrum has been termed the 'water hole'.

NASA has proposed a major SETI project to cover the whole sky at a wide range of wavelengths, with particular emphasis on the water hole. The search would use existing radio telescopes connected to special receivers designed to pick out faint alien signals. The whole project is intended to last five years, and should get under way in the 1980s if funding is forthcoming.

Message to the Stars

What sort of message might we expect to receive from the stars? An example is the signal sent in 1974 from the Arecibo radio telescope in Puerto Rico to a distant star cluster. Arecibo is the world's largest radio telescope, 305 m (1000 ft) in diameter. It is so powerful that it could communicate with an identical twin on the other side of the Galaxy, the best part of 100 000 light years away.

Therefore we already have the ability to swap signals with aliens of equivalent technological level to ourselves–if only we can find them! The Arecibo message consisted of 1679 on-off pulses. If these pulses are arranged into a grid 23 characters wide by 73 deep, they form a dot picture telling aliens about our body chemistry, our appearance and size and our solar system.

The Arecibo message was only a demonstration broadcast, not a serious attempt at interstellar signalling. It lasted a mere three minutes, and was sent only once, in the direction of a cluster of 100 000 stars called M13 in the constellation Hercules. So far

Left: Inside Arecibo's control room. *Above:* Part of the Arecibo radio telescope's receiving equipment.

away is the cluster that even if any aliens there happened by chance to pick up the transmission and replied immediately, we would not hear from them until about AD 50 000.

A major hindrance to communication between alien civilizations is the language problem. Sending a dot picture overcomes this as considerable information can be conveyed in pictorial form. However, the message could be a lot more complex than this. Computers converse in the on-off pulses known as binary code, and possibly vast amounts of information will come streaming in over the interstellar radio link that will require our own computers to decode. Whatever the case, the prospects are fascinating–and not a little scaring.

Cyclops–Space's 'Big Ear'

But what if the incoming signals are too weak to be detected by existing radio telescopes–or if the aliens are not deliberately signalling to us? To increase sensitivity, we need to build much bigger collecting arrays. One fantastic scheme is for a big ear in space called Cyclops, built up from as many as 1000 radio dishes, each 100 m (109.3 yds) in diameter. Cyclops would cover an area of 20 square km (7.7 square miles), and cost an estimated 10 000 million dollars. There seems little chance of it actually being built. Better might be a giant antenna in space, built by the same orbiting industrial facilities as are expected to be producing power satellites and space colonies next century.

Such large collecting areas would be able to eavesdrop on the aliens' radio and TV transmissions leaking out from their home planet. And, even though we are not sending deliberate radio messages to the stars ourselves, alien civilizations would be able to eavesdrop on our domestic communications in just the same

Artist's impression of an aerial view of a lunar Cyclops interstellar search system.

A portion of the terrestrial Cyclops system antenna array.

way. For instance, TV and radio broadcasts from Earth could be detected at a distance of 25 light years, a range encompassing about 300 stars. Our most powerful transmissions come from early-warning radars which, in time, could be detected at distances of 250 light years, a range encompassing several hundred thousand stars. Therefore, aliens could deduce a great deal about life on our politically divided planet. Inhabitants of Alpha Centauri and points beyond would not be glued to their screens for our TV series because programme information itself would be too weak to decipher over interstellar distances. But aliens could detect the powerful carrier beams on which the programmes ride.

Greetings from Earth

Some people think that space probes are a good way to send messages to the stars, despite the fact that they travel much more slowly than light or radio waves. We have already begun to send messages to the stars by space probe, in the form of the plaques attached to Pioneer 10 and 11, and the long-playing records carried by the two Voyager probes to the outer planets. Eventually, all four probes will leave the solar system and drift out among the stars of the Galaxy, where they may be intercepted by some advanced civilization millions of years from now.

Each of the Pioneer plaques consists of simple engravings showing the position of our Sun in relation to several pulsars; a schematic diagram of our solar system; and two human beings. The Pioneer plaques are like a cosmic greeting card. By contrast, the information carried aboard the Voyager probes is altogether richer and more complex. Into the grooves of a long-playing record are encoded written messages, pictures of our planet, spoken greetings, various sounds of life on our planet, and a musical selection of Earth's greatest hits, ranging from the music of primitive societies to Bach, Beethoven, and even Chuck Berry. These records, if they are ever found, would doubtless give aliens much cause for head-scratching, or whatever their equivalent of head-scratching is.

One day, of course, we might detect some similar interstellar 'message in a bottle' floating into our own solar system, although the chances are slight.

More likely, during our explorations of the solar system we will

Greetings from Earth: the plaques attached to Pioneer 10 and 11.

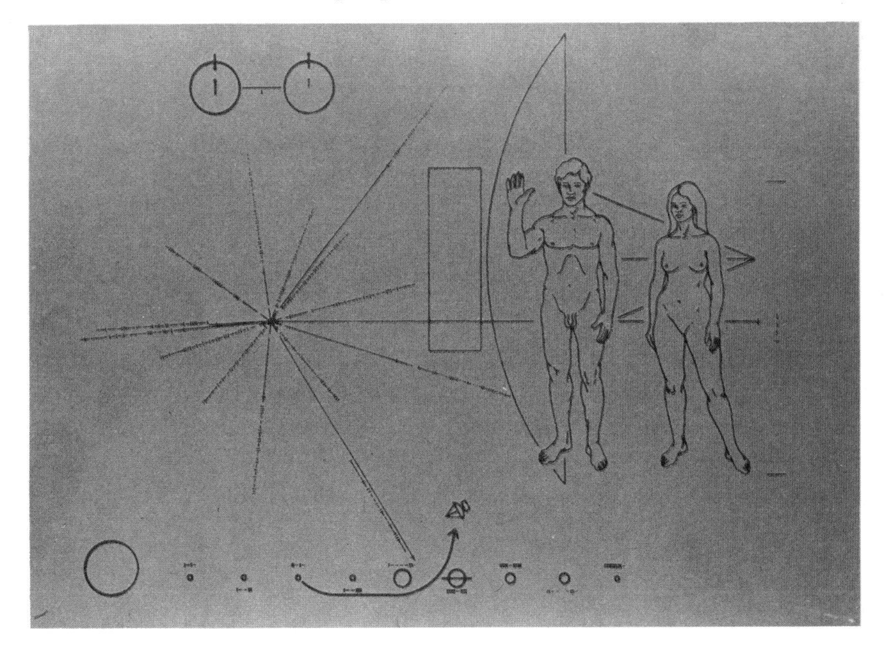

encounter abandoned artefacts of an alien expedition, including, perhaps, a commemorative plaque similar to that attached to the Apollo 11 lunar module. Despite misguided claims, no evidence has yet been uncovered that Earth was visited in the past by aliens.

Most interesting of all would be a probe controlled by a powerful computer programmed to communicate with any civilizations it found. Such probes could be sent to all likely stars and, on arrival, might park themselves in orbit to listen for signs of activity on the planets below. When radio noise from an advanced civilization was detected, the probe would attempt to make itself known by sending out its own signals. We should listen out in case there is such a probe in the solar system at present.

The Aliens Are Here!

According to some people, we need not waste our time listening for extraterrestrial radio messages because the aliens are already here. Alien spacecraft are said to have been seen by many people in the form of UFOs, or flying saucers–and it is even claimed that certain humans have met the occupants of these alleged craft. If these astounding stories are true, they would clearly be of major importance to science. So why do most scientists reject the UFO field? Is there any real evidence that UFOs are extraterrestrial visitors?

Stories of strange objects in the skies are to be found in ancient texts, including the Bible, but it seems probable that most, if not all, of these were misinterpretations and over-dramatizations of natural phenomena, or even pure fantasies and inventions.

The modern flying saucer era began in 1947 with the report by Kenneth Arnold, a private pilot, of a sighting of nine bright objects flying like saucers skipping over water, as he described them. Since then, a growing torrent of stories of strange lights, apparently inexplicable happenings, and even landings and abductions by aliens has flooded in, aided by media only too ready to publicize a sensational story.

Could aliens like these exist, or are they just a science-fiction dream?
Opposite, above: Sightings and photographs of UFOs pour in from all over the world, like this picture taken in Austria. One of the spotters said he'. . . . saw a disc form moving about and up and down like a falling leaf . . . Sometimes I believed the object would land, so low it came down. . . .'
Opposite, below: An alleged UFO photographed by a group of Dutchmen in southern Switzerland. Most scientists dismiss such UFO pictures as misinterpretations of natural objects, or fabrications.

Identified Flying Objects

The vast majority of UFO reports–90 per cent or more–can quickly be explained as misidentifications of natural or manmade objects. The main offenders are bright planets (particularly Venus) and stars, aircraft, meteors, and satellites. It is usually possible to identify these objects conclusively because the reports are reliable, and the same object is often seen by several persons at the same time. But the most baffling reports tend to come from isolated individuals or small groups. These reports include alleged meetings with extraterrestrials or even abduction and medical examination by them. Here we are dependent on the witnesses' reliability and honesty.

We ask whether it is really likely that an alien expedition would cross countless light years of space, successfully evade the air defences of the entire world, and then briefly land beside a remote roadside to snatch two passing motorists (one male and one female, naturally) for a brief examination, before departing as invisibly as they arrived?

By contrast, when something unexpected but undeniably real does pass through the atmosphere, such as a meteorite or reentering satellite, there is no shortage of evidence. For instance, in 1972 a brilliant fireball streaked through the atmosphere over the Rocky Mountains before

returning to space. This event occurred over one of the least populated areas of North America, but the object was seen by thousands (including scientists), photographed by several, and even tracked by a US Air Force surveillance satellite.

At a time when Earth and its surroundings are kept under closer scrutiny than ever before by military and civilian agencies, and with so many camera-happy tourists combing the globe, it is inconceivable that genuine extraterrestrial visitors could arrive and depart without leaving some firm evidence of their visit. It need hardly be said that not one UFO has ever been identified as an extraterrestrial spacecraft.

Of course, we should never give up hope that one day we may discover the genuine article. But, as of now, it has to be admitted that we don't know for sure of the existence of life anywhere else in the Universe other than on Earth.

Artist's impression of a scene that didn't happen on the Moon — Apollo astronauts are captured by alien beings. But what of the future . . .?

GLOSSARY

(Note: Italicized words appear as separate entries.)

Apogee The closest point to Earth of a *satellite* in its *orbit*.

Apollo Project The American space programme to place men on the *Moon*. The Apollo spacecraft contained three men, two of whom entered the *Lunar Module* for making a Moon landing. Apollo spacecraft were launched by *Saturn* rockets. As well as Moon missions, Apollo spacecraft were used to take astronauts to the *Skylab space station*, and also for a joint docking mission in orbit with a Soviet *Soyuz*.

Ariane A rocket built by the *European Space Agency* for launching *satellites*.

Ariel satellites A series of six UK satellites, launched by American rockets, which started in 1962. The first four studied radio astronomy and Earth's upper atmosphere. Ariel 5 studied *X-ray radiation* coming from the sky. Ariel 6 was launched in June 1979, to study cosmic rays and X-rays.

Asteroid A small body orbiting the *Sun*, also known as a minor *planet*. Most *orbit* in a belt between Mars and Jupiter. Over 2000 asteroids are known, ranging in diameter from 1000 km (620 miles) to a few km (or miles).

Astronomical unit The average distance of Earth from the *Sun*, 150 million km (93 million miles).

Atlas An American rocket used to launch *satellites* and space probes. Astronauts in the Mercury series were orbited by Atlas rockets.

Big Bang The explosion which is believed to have marked the origin of the *Universe* as we know it, between 10 000 million and 20 000 million years ago. The Universe has been expanding since the Big Bang.

Black Hole A region of space where gravity is so intense that nothing can escape, not even light. Black holes are believed to form as a *star* collapses under the inward pull of its own gravity at the end of its life.

Cape Canaveral The main American space launching site, in Florida. All manned American space launches, and most *satellites* and space probes, depart from Cape Canaveral.

Comet A body made of rock and gas which *orbits* the *Sun*. Most comets have very elongated (elliptical) orbits. As they approach the Sun, the gases evaporate to form a long, flowing tail.

Communications Satellites Satellites which relay telephone and television signals around the world. Most communications satellites are in so-called *geostationary orbit*. The main series of communications satellites is *Intelsat*.

Constellation A pattern of *stars*, often visualized to form the figure of a legendary hero or mythical animal. A total of 88 constellations cover the sky.

Cosmos Satellites A series of Russian Earth satellites, for scientific and military purposes and for tests of new space probes or manned craft. Over 1000 Cosmos satellites were launched between 1962 and 1979.

Crab nebula An irregular-shaped patch of gas in the *constellation* of Taurus, it is the remains of a *star* seen to explode as a *supernova* in 1054 AD. At its centre is the fastest-flashing *pulsar*.

Delta An American rocket for launching scientific and *communications satellites*.

Eclipse The shadowing of one body by another. When the *Moon* enters Earth's shadow, a lunar eclipse is seen. At an eclipse of the *Sun*, the Moon passes across the face of the Sun, temporarily blotting out its light.

Escape Velocity The minimum speed at which a rocket must travel to break free from the grip of a planet's gravity. Escape velocity from Earth is 11·2 km (7 miles) per second; from the *Moon* it is 2·4 km (1·8 miles) per second.

European Space Agency An organization of European countries to build and launch *satellites* for space research.

EVA Abbreviation for extravehicular activity, commonly known as a space walk.

Explorer satellites Series of American scientific satellites. The first American satellite was Explorer 1, launched on January 31, 1958.

Galaxy A mass of *stars* bound together by gravity. The smallest galaxies contain a few million stars; the largest contain more than a million million stars. Our own Galaxy is also known as the *Milky Way*.

Gemini Project A series of American space missions. Astronauts in the two-man Gemini spacecraft made space dockings and set new records for space walks and long-duration flights.

Geostationary Orbit An orbit at an altitude of 35 900 km (22 306 miles) in which a *satellite* revolves around Earth every 24 hours, the same rate at which Earth spins. Therefore the satellite appears to hang stationary over one spot on the equator. It is also known as a synchronous orbit.

Heat Shield A protective layer on a spacecraft to insulate it from excess heat while re-entering Earth's atmosphere.

Intelsat A series of *communications satellites*, built and launched by the International Telecommunications Satellite Corporation. The first of the series was Early Bird, in 1965. The latest generation of satellites, Intelsat 5, carries 12 000 telephone channels.

Landsat A series of three Earth-survey *satellites* that monitored crop growth, air and water pollution, and searched for possible new sources of minerals.

Launch Window The time during which a *satellite* or probe can be successfully launched, depending, for instance, on the position of the *Moon*, a *planet*, or another spacecraft in *orbit*.

Light Year The distance travelled by a beam of light in one year, equivalent to 9·5 million million km (5·9 million million miles).

Luna Probes Series of Russian

Moon probes, which have landed on the *Moon* and have studied it from *orbit*.

Lunar Module The craft used by astronauts for landing on the *Moon* during the *Apollo* programme.

Lunar Orbiter A series of five American probes which surveyed the *Moon* from orbit to spy out suitable landing sites for astronauts.

Lunokhod Unmanned Russian Moon car which roamed the lunar surface under radio control from Earth.

Magnetosphere The magnetic shell caused by the extension of Earth's magnetic field into space.

Mariner Spacecraft Series of American probes to the *planets*. Mariner craft have visited Venus, Mars and Mercury.

Mercury Project America's first manned space programme. Astronauts *orbited* Earth in the single-man Mercury capsule, proving that a human could successfully fly in space and return safely.

Meteor A tiny particle from space which enters Earth's atmosphere, burning up to form a brilliant shooting star. A typical meteor is about the size of a grain of sand. Meteors are believed to be dust from *comets*.

Meteorite A lump of rock or metal from space which plunges through the atmosphere to hit the ground. Meteorites are believed to be fragments from *asteroids*. A meteorite which hits the ground at a sufficiently high speed will dig a giant crater, like those on the *Moon* and *planets*.

Moon Earth's natural *satellite*, 384 000 km (238 000 miles) away and 3480 km (2160 miles) in diameter. Natural satellites of other planets are also termed moons.

Milky Way The Galaxy of at least 100 000 million stars of which the *Sun* is one. The Milky Way is spiral in shape, like a catherine-wheel, is about 100 000 light years in diameter, and the Sun lies about 30 000 light years from the centre.

NASA The National Aeronautics and Space Administration, a US government body for peaceful exploration of space.

Nebula A cloud of dust and gas in the *Galaxy*. Some nebulae are lit up by the light from *stars*, such as the Orion nebula. Other nebulae are dark, and can only be seen in silhouette against the star background.

Neutron Star A small, dense star in which the protons and electrons of the star's atoms have been crushed together to form the atomic particles known as neutrons. A typical neutron star contains as much mass as the *Sun* compressed into a ball only 20 km (12·4 miles) across. Neutron stars are believed to be formed in *supernova* explosions.

Orbit The path in space of one body around another. Orbits are usually elliptical in shape (like squashed circles), as a perfectly circular orbit is virtually impossible to attain.

Perigee The furthest point from Earth a *satellite* reaches in its *orbit*.

Pioneer A series of American solar system probes. Pioneer spacecraft have investigated conditions in the space between the *planets*, as well as visiting Jupiter, Saturn and Venus.

Planet A non-luminous body which shines by reflecting the *Sun*'s light. The planets of our solar system are made from rock, metal, and gas. Other *stars* may also have planets.

Pulsar A rapidly flashing radio source. Pulsars are believed to be small *stars* (probably *neutron stars*) spinning quickly, giving out a flash of radiation like a lighthouse beam each time they turn. The fastest known pulsar flashes 30 times a second, the slowest every 4 seconds.

Quasar An intensely luminous object far off in the *Universe*. Most astronomers think quasars are the centres of massive *galaxies* seen at an early stage in their evolution. Quasars may be powered by hot gas falling into giant *black holes*.

Radio Telescope A device for collecting radio waves from space. For instance, hydrogen gas emits radiation naturally at a wavelength of 21 cm. Most radio telescopes are shaped like dishes and operate in a similar way to an optical reflecting telescope. Similar devices to radio telescopes are used for tracking *satellites* and space probes.

Ranger A series of American *Moon* probes which sent back the first close-up photographs of the Moon, in 1964 and 1965.

Red Dwarf A *star* smaller and cooler than the *Sun*.

Red Giant A *star* much larger and brighter than the *Sun*. Stars swell up into red giants at the ends of their lives. Red giants can be 100 million km (62 million miles) or more in diameter.

Re-entry The return of a spacecraft into Earth's atmosphere. Most spacecraft burn up by friction with the atmosphere unless they are specially insulated with a *heat shield*.

Retro-rocket A small rocket or group of rockets aboard a spacecraft which slow down its motion, for instance to return from *orbit* to Earth.

Salyut A series of Soviet *space stations*, made from the converted top stage of a Proton rocket. Salyut is roughly cylindrical in form, 12 m (39 ft) long and up to 4 m (13 ft) in diameter. The first Salyut was launched in 1971. Salyut 6 was launched in 1977.

Satellite An object which *orbits* around another. All the planets except Mercury and Venus have natural satellites (moons). The world's first artificial satellite was *Sputnik*.

Saturn Rocket A family of rockets developed specifically for the American manned space programme. Saturn 1B, the smaller version, could launch a three-man *Apollo* capsule into orbit around Earth. For *Moon* missions, and to launch the *Skylab* space station, the larger Saturn V, the most powerful rocket yet developed, was used. Saturn rockets are no longer used.

Scout A small American solid-fuelled rocket, used for launching scientific *satellites*.

Service Module The section of a manned spacecraft behind the crew compartment containing supplies such as air, water, and electricity.

Skylab An American manned *space station*, launched in 1973, made from the top stage of a *Saturn V* rocket. Three crews, each of three men, occupied Skylab in 1973–4. The abandoned *space station* burned up in the atmosphere in 1979. Skylab was the largest object ever put into *orbit*, weighing 75 tonnes and measuring over 25 m (82 ft) long.

Solar Cell A device for turning sunlight into electricity. Most *satellites* and probes get their power from solar cells. Solar cells are mounted on the outside surface of the spacecraft, or on long solar panels.

Solar System The collection of nine *planets*, their *moons*, and various smaller objects such as *asteroids* and *comets*, all of which are orbiting the *Sun*.

Soyuz A Soviet manned spacecraft, introduced in 1967. Soyuz is used to ferry two-man crews up to *Salyut space stations*, although it has also been used to make independent missions in *orbit* around Earth with up to three cosmonauts on board.

Space Shuttle The winged, re-usable space plane which is replacing conventional rockets for many purposes. The Shuttle, which is launched like a normal rocket, is piloted into *orbit* by astronauts, carrying a payload in its cargo bay. It can also bring objects back from orbit in its cargo bay, such as malfunctioning or disused *satellites*. The Shuttle glides back to land on a runway, like an aircraft. It can then be refuelled and relaunched, thereby making spaceflight much cheaper.

Space Station An enclosed structure in which astronauts can live and work for long periods. Early space stations were *Salyut* and *Skylab*. A small space station called *Spacelab* has been built to fly with the *Space Shuttle*.

Spacelab A small *space station* built by the *European Space Agency*, which will be carried into *orbit* and return to Earth in the cargo bay of the *Space Shuttle*. American and European astronauts will work in Spacelab for a week or more at a time.

Sputnik A series of Soviet Earth *satellites*. Sputnik 1, launched on October 4, 1957, was the world's first artificial satellite. Sputnik 2, a month later, carried the first living thing into *orbit*, the dog Laika. The series ended in 1961 with a series of test flights of an unmanned *Vostok*.

Star A glowing ball of gas. Stars produce their own heat and light from nuclear reactions at their centres. Stars consist of about 80 per cent hydrogen, the most abundant element in the *Universe*, almost all the rest being made of helium.

Steady State A theory which says the *Universe* had no beginning and will never end, but has always looked much the same. Most astronomers now discredit the theory in favour of the *Big Bang*.

Sub-orbital A space launch which does not go into orbit around Earth, but which raises a payload to the edge of the atmosphere before dropping back again. The first two American manned launches of the *Mercury* programme were sub-orbital.

Sun Our parent star, an incandescent ball of gas 1·4 million km (865 000 miles) in diameter. The Sun is average in size and brightness, but appears much more prominent to us than the night-time *stars* because it is so much closer.

Sunspot A dark patch on the *Sun*'s surface, caused by a cooler region of gas. The number of sunspots waxes and wanes every 11 years or so, a period termed the solar cycle.

Supernova The explosion of a *star* several times more massive than the *Sun* at the end of its life. The shattered outer regions of the star are thrown off into space to form an object such as the *Crab nebula*. The core of the erupted star is left behind as a tiny *neutron star*, or even a *black hole*.

Surveyor A series of American probes which landed on the *Moon*'s surface in 1966–8, showing that it was safe for men to follow.

Telescope An optical device for collecting light. Telescopes can collect light by means of a lens (a refracting telescope) or by a mirror (a reflecting telescope). Since a large lens or mirror collects more light than the human eye, telescopes can see objects much fainter than can the eye. They can also make objects seem nearer by magnifying the image. The largest telescopes are all of the reflecting variety. Some telescopes collect invisible radio waves from space. These are called *radio telescopes*.

Titan rocket A family of American space launchers. Titan rockets of various design have been used to launch *satellites*, manned *Gemini* spacecraft, and, with extra boosters added, probes such as *Viking* and *Voyager* to the planets.

Universe Everything around us. The Universe is dotted with *galaxies* of *stars* as far as the largest *telescope* can see. The Universe is expanding, apparently from its origin in a massive explosion (the *big bang*) between 10 000 million and 20 000 million years ago.

Viking Two American space probes to Mars. One half of each Viking surveyed the planet from *orbit*, while the other half landed on the surface. They didn't find life.

Voskhod Soviet manned spacecraft, modified from *Vostok*. Voskhod 1 in 1964 carried the first three-man crew into *orbit*. From Voskhod 2, in 1965, Alexei Leonov made the first space walk.

Vostok Soviet manned spacecraft which carried the first human, Yuri Gagarin, into *orbit* on April 12, 1961. A total of six Vostok flights were made by different cosmonauts.

Voyager Two American space probes to the outer planets. The two Voyagers reached Jupiter in 1979 before flying on to Saturn. Eventually, one Voyager may reach Uranus and even Neptune, though this won't happen until 1986 and 1989 respectively.

Weather Satellites Several series of Earth satellites which photograph clouds and measure the temperature and moisture content of the atmosphere, allowing meteorologists to make more accurate forecasts. The first series of weather satellites was called Tiros, which began in 1960. Modern weather satellites are often put into *geostationary orbit*, such as the European Meteosat, for example.

Weightlessness A condition experienced by objects in space, also known as free fall. In space, an object is falling with nothing to support it, and so becomes weightless. This applies to spacecraft and all the objects in them. Weightlessness is sometimes incorrectly termed zero-gravity.

White Dwarf A small, hot *star* about the size of Earth. White dwarfs are believed to be the end point of the evolution of stars like the *Sun*.

X-ray astronomy The study of X-ray radiation coming from space. X-rays are short-wavelength radiation emitted by hot gas, with a temperature of many millions of degrees. Such high temperatures are produced in the gas thrown out from a *supernova* explosion, or in gas falling on to a *neutron star* or *black hole*.

INDEX

Numbers in bold refer to illustrations